U.S. Department
of Transportation
Research and Special
Programs Administration

DOT-VNTSC-FAA-01-05

April 23, 2001

Traffic Situation Display Familiarization For Controllers
Final Report

Prepared for:

Federal Aviation Administration
Free Flight Phase 1 Program Office
1500 K St. N.W., Suite 500
Washington, D.C. 20005

Prepared by:

Office of Operations and Traffic Management
U.S. Department of Transportation
Research and Special Programs Administration
Volpe National Transportation Systems Center
Cambridge, MA 02142-1093

Acknowledgments

The TSD Familiarization for Controllers Final Report was prepared under the project management of Dr. Sylvia A. Harris, U.S. Department of Transportation, Volpe National Transportation Systems Center (Volpe Center), Telecommunications Division. It was written by Dr. Sylvia A. Harris, Telecommunications Division, Olive LeSueur, Intermodal Logistics Systems Planning and Integration Division, and Joyce Westner and Joseph Jankowski, Brattle Systems, Inc.

Table of Contents

Background 1
   General Information 1
   Pre-Travel Preparation 1
   Familiarization Sessions 1
   Evaluation/Feedback 2
   Hardware/Software 3
   TSD Trainers 3

Lessons Learned 4
   Recurring Themes from Special Factors 4
   Recurring Themes in the Evaluation Comments 4

Conclusions/Recommendations 5
   Conclusions 5
   Recommendations 5

Appendix A: Preparing For TSD Training at ARTCCs A-1

Appendix B: Summary TSD Training Schedule B-1

Appendix C: TSD 2000 Coaching Exercises C-1

Appendix D: TSD 2000 Coaching Exercises – Instructor Version D-1

Appendix E: TSD For Controllers Coaching Exercise Evaluation E-1

Appendix F: TSD Familiarization Evaluation Reports F-1

Appendix G: TSD Familiarization Training Issues G-1

Appendix H: TSD 2000 Trainer Certification Checklist H-1

Appendix I: Job Aid I-1

BACKGROUND

General Information

The Volpe National Transportation Systems Center's (Volpe Center) Automation Applications Division (DTS-56) is developing and implementing the FAA's Enhanced Traffic Management System (ETMS) as one component of ongoing research and development to improve FAA Air Traffic Control through the Advanced Traffic Management System (ATMS). DTS-56 has requested support for ETMS in the design, development and delivery of training.

Based on an initial analysis of training needs, the Volpe Center developed 2 hours of computer-based instruction (CBI) for the Traffic Situation Display (TSD) on the FAA's computer-managed instruction (CMI) Compliant workstation and a supplementary "field site training program" to provide traffic managers with one-on-one training, on location, immediately after TSD was installed at each Traffic Management facility.

The FAA requested that a similar TSD familiarization program be developed and delivered to Air Traffic Control (ATC) Supervisors and Controllers-in-Charge. This program was designed to insure that each user of this mission-critical system has the individual coaching and support he/she may need to display TSD information, and that all program features which are of benefit to the Air Traffic Control community were introduced to each user.

The audience for the familiarization program was ATC supervisors and CICs at all 21 FAA Air Route Traffic Control Centers (ARTCCs).

Pre-Travel Preparation

For each training team, one of the trainers served as the Lead Trainer. In preparing for a site visit the Lead Trainer would use "Preparing For TSD Training at ARTCCs" (Appendix A) to ensure key items were completed, including coordinating with the site point-of-contact (POC) a training schedule and the number of trainees. The success of the familiarization often depended upon this coordination between the Lead Trainer and the site POC. At most sites, the POCs were available and able to schedule students, arrange access for the instructors, and help the training run smoothly.

Familiarization Sessions

TSD familiarization was conducted at seventeen sites (Appendix B), with the training team consisting of a Lead Trainer and one or more additional trainers depending on the site (large sites had three trainers, small sites two). Fifteen out of 21 ARTCCs were visited (six declined). TSD Familiarization was also conducted twice at the Command Center. A total of 665 TSD users attended the sessions.

Familiarization was conducted in small groups consisting of a trainer and no more than three students. After a short demonstration by the trainer, the students took turns at the keyboard while the trainer coached them through the software. Each of these sessions lasted two hours; each trainer was available for three sessions per day. Trainers were extremely flexible in meeting the needs of the sites. At most sites, the trainers worked shifts to cover both day and evening hours. On one occasion, a trainer even conducted a 4:00 am session. Each trainee received training per the TSD Coaching Exercises (Appendix C). An Instructor Version of the Coaching Exercises (Appendix D) was also developed to insure consistent quality. At the completion of each session, the trainee filled out an evaluation (Appendix E) and was issued a TSD Job Aid (Appendix I). The training materials, which the POC trainer brought to each site, included:

- TSD Coaching Exercises
- TSD Coaching Exercise Instructor's Version
- TSD Coaching Exercise Evaluation
- Enhanced Traffic Management System (ETMS) Reference Manual

Evaluation/Feedback

At the completion of training, the POC trainer prepared an Evaluation Report (Appendix F) for that site. The report included the trainee evaluation feedback, any special factors or problems, and specific comments from trainees. The feedback was used to modify training as needed.

Each trainee rated four specific statements; the statements and the overall results for the seventy-three sites are shown below.

1. My ability to use TSD has increased.
   | 1 | 2 | 3 | 4 | 5 |
   | Strongly Disagree | Disagree | Neutral | Agree | Strongly Agree |

2. The instructor presented information clearly and understandably.
   | 1 | 2 | 3 | 4 | 5 |
   | Strongly Disagree | Disagree | Neutral | Agree | Strongly Agree |

3. The instructor appeared knowledgeable about the TSD software.
   | 1 | 2 | 3 | 4 | 5 |
   | Strongly Disagree | Disagree | Neutral | Agree | Strongly Agree |

4. This training will help me in my job.
   | 1 | 2 | 3 | 4 | 5 |
   | Strongly Disagree | Disagree | Neutral | Agree | Strongly Agree |

Overall Results:

Question 1	Question 2	Question 3	Question 4
4.60	4.79	4.81	4.54

April 23, 2001

Hardware/Software

The ETMS software runs on a network of Hewlett-Packard computers. These computers run the HP-UX operating system, which is a version of UNIX. ATC Supervisors and CICs access the system using a Windows-based PC running Exceed software, an X-Windows interface. ETMS and TSD are then displayed in an Exceed window.

TSD Trainers

The TSD trainers were responsible for familiarizing all available ATC Supervisors and CICs with the TSD software. Prior to conducting familiarization, the trainers completed 160 hours of preparation time and obtained hands-on experience at either the Volpe Center in Cambridge, MA, the Command Center in Herndon, VA, or the TSD lab facility at the FAA Academy in Oklahoma City, OK. At the end of their training period, the trainers were certified on the system using the TSD Trainer Certification Checklist (Appendix H). In total, eight trainers participated in this project, 4 of which participated in the previous TSD training. Background materials for the instructors included:

- Preparing For TSD Training at ARTCCs
- TSD Trainer Certification Checklist
- Using TSD at the Volpe Center
- TSD Glossary
- Traffic Situation Display Acronyms
- Glossary of FAA Acronyms
- *Fundamentals of Air Traffic Control* by Michael S. Nolan
- AOPA's "Airspace for Everyone"
- Web Sites of Interest
- Lessons Learned From the First TSD Roadshow
- ETMS Reference Manual
- ETMS Tutorial

LESSONS LEARNED

Recurring Themes from Special Factors

The Special Factors sections from the TSD Familiarization Evaluation Reports (Appendix F) show some general themes regarding problems or issues that affected training. Overall, the training went more smoothly. Some common themes gathered from these sections follow.

- The location of the workstations was less than optimal. Noise was sited several times, either from loud equipment nearby or other training/telephone conferences in close proximity. At two sites, workstations were too close together, making it difficult when more than one session was conducted at the same time.

- At some sites, students came to training with the idea that they were receiving training on the ESIS projection system and/or Exceed. Many of these students left the TSD training disappointed.

- Three site POCs (at ZLC, ZJX, and ZFW) were singled out for having an excellent plan and high turnout.

- Training on the ESIS projection and Exceed was given for 30 minutes prior to the TSD training at two sites, ZLC and ZJX.

Recurring Themes in the Evaluation Comments

The comments sections from the TSD Familiarization Evaluation Reports (Appendix F) reveal some common opinions that the students had about the TSD training. The themes collected from the students' comments follow.

- The students made many positive comments about:
    - hands-on training
    - small groups of trainees
    - easy to use software
    - individual pacing and focus of the training
    - ability to ask questions and get answers immediately
    - praise for specific software functions such as flights, weather, etc.
    - praise for instructors

- The students thought the following areas of the training could use improvement:
    - need for recurring training, or training a few weeks after having a chance to use system
    - more hands-on/practice time
    - include training on the ESIS projection system and/or Exceed

CONCLUSIONS/RECOMMENDATIONS

Conclusions

The training development team, based on the site evaluation reports and the trainers' experiences, draws the following conclusions.

- Early communication and contact with the site POC was essential to the success of training at a site. In addition, the availability of the site POC for continuous schedule revisions and updates was helpful.

- Good site coordinators were the key to successful training.

- The CBI made a positive difference at sites where students were required to view it prior to training.

- The flexibility of the instructors helped assure successful results in many special situations.

- The audience universally appreciated hands-on, small group training.

Recommendations

Based on the site evaluation reports and the students' comments, the training development team recommends the following methods for future training projects.

- After the controllers have used TSD for a while, offer follow-up training.

- Clarify the training to be given in advance. A letter to the site POC explaining what will be covered is recommended. In addition, consider including with this letter a flyer for posting on bulletin boards announcing the timing and content of training.

- Create a recovery plan in case the system goes offline.

- Optimize the physical location of the training workstations to avoid noise and overcrowding.

April 23, 2001

… Appendix A: Preparing For TSD Training at ARTCCs

Preparing For TSD Training at ARTCCs

Table of Contents

At Least three weeks before the class (Contract Trainers)	A-2
At Least three weeks before the class (All Trainers)	A-2
Two weeks before the class (Olive Lesueur)	A-2
The Generic Schedule	A-3
The week before arrival (Lead Trainer)	A-3
Bring to the site (Lead Trainer)	A-3
Bring to the site (All Trainers)	A-4
The day of arrival on-site (Lead Trainer)	A-4
The first day of training (First Shift Trainers)	A-4
The first day of training (Second Shift Trainers)	A-5
During coaching sessions (All Trainers)	A-5
At the end of each coaching session (All Trainers)	A-6
At the completion of training (Lead Trainer)	A-6
Monthly teleconferences (All Trainers)	A-6
Training Sites	A-7
Contacts	A-8

April 23, 2001

At Least three weeks before the class (Contract Trainers)

- Submit a travel request to the Brattle Travel Coordinator, cc Joe.

- If you want to purchase your ticket, wait until your travel request is approved.

At Least three weeks before the class (All Trainers)

- Plan to arrive on Monday and return on Friday to allow for maximum scheduling flexibility.

- The Lead Trainer should arrive early enough to coordinate with the POC by visit or telephone. The Lead should have a meeting with the team early evening to disseminate the latest training particulars. All trainers should plan to arrive early enough on Monday to attend that meeting.

Two weeks before the class (Olive Lesueur)

- Call Joe to coordinate the following:
    - Site:
    - Site Coordinator:
    - Training Dates:
    - Lead Trainer:
    - Co-trainer(s):

- Call the site coordinator to determine:
    - Estimated number of students (max of 3 per session, max of 3 sessions per trainer per day, max of 3 days)
    - Proposed Schedule (see The Generic Schedule, page 3)
    - Number of TSD terminals or PCs (two should be the minimum)

Email Joe with the site information. Joe will forward to trainers.

The Generic Schedule

- For Large ARTCCs:
    - Tuesday: 2 day shifts and 1 evening shift
    - Wednesday: 2 day shifts and 1 evening shift
    - Thursday: 1 day shift and 2 evening shifts
    - 2 rental cars required for the 3 trainers

- For Small ARTCCs:
    - Tuesday: 2 day shifts
    - Wednesday: 2 midday shifts
    - Thursday: 2 evening shifts
    - 1 rental car required for the 2 trainers

- For All ARTCCs:
    - Each trainer will coach up to 3 trainees per session
    - Each trainer will be assigned no more than 3 sessions per day
    - No trainer will work an evening shift one day and a morning shift the next

The week before arrival (Lead Trainer)

- Call the site coordinator to discuss site logistics:
    - Number of TSD terminals or PCs available for training (should be at least two)
    - Schedule (suggest the Generic Schedule)
    - Directions to site and any special instructions on gaining access (IDs, parking)
    - Who to meet when you first arrive

- Call or email your co-trainer(s) to coordinate travel and training schedules.

- Gather materials to bring to the site:
    - Evaluation forms
    - Coaching Exercise forms
    - Job Aids

Bring to the site (Lead Trainer)

- Schedule
- Evaluation forms for all students
- Coaching Exercises for all students
- Job Aids for all students

Bring to the site (All Trainers)

- ETMS Reference Manual
- Preparing For TSD Training at ARTCCs (this document)

The day of arrival on-site (Lead Trainer)

- Call the site coordinator to update logistics.
- Meet with the team.

The day of arrival on-site (All Trainers)

- Attend the early evening team meeting.

The first day of training (First Shift Trainers)

- Meet the site coordinator to finalize site logistics:
  - Schedule
  - TSD terminal locations
  - User ID and password
  - Location and phone number of telephone for trainer use
  - Location of break room, bathrooms, etc.

- Perform on-site technical preparation on each workstation as early as possible to leave time for fixing any problems. Verify the default display has the settings below. If not, create and save a Training display and recall it during you coaching sessions.
  - Display the TSD window on the top half of the screen.
  - Display the Times window.
  - Display a black Background.
  - Display gray Boundaries.
  - Display brick red ARTCC boundaries.
  - In the Select Flights window, define arrival sets at ORD, DFW, and ATL, but do not check the Show box. Hide Flights.
  - In the Select Alerts window, check Airports All. Hide alerts.
  - In the Select Weather window, check NOWRAD 2 km. Hide weather.
  - Create a private reroute (ask for a commonly used reroute within the center). Hide Reroutes.

The first day of training (Second Shift Trainers)

- Coordinate with the first shift trainers to learn final site logistics:
  - Schedule
  - TSD terminal locations
  - Location and phone number of telephone for trainer use
  - Location of break room, bathrooms, etc.
  - Lessons learned during the first shift

During coaching sessions (All Trainers)

- Arrive on time for each session. Finish each session on time.

- Explain to the students what you will be demonstrating, and what they will do.

- Follow the Coaching Exercises – Instructor Version document.

- If you have TSD or hardware problems, call the network administrator or the Help Desk. Note any problems on the back of the Coaching Exercises document. Be specific. This will be used in the training report.

- If the PC/Exceed system goes down, ask the coordinator if there are ETMS workstations available for training.

- Allow students to ask questions throughout.

- Allow students to work independently during Part 4, but remain available to answer questions.

- Do not discuss government, FAA, or union policies with the students; never discuss Volpe/Brattle relations.

- Project a positive image and keep a pleasant attitude. Dress professionally at all times. Remember how important it is to be positive about the training and the software; you are "marketing" the Volpe team and the software.

- If questions about the TSD come up that you can't answer, call Joe or Joyce. Tell the student you will get back to them and do so.

- Have fun!

April 23, 2001

At the end of each coaching session (All Trainers)

- Ask students if they have any questions.

- Have each student complete the session evaluation form.

- Complete a TSD Coaching Exercise document (*one per session*) and give to the Lead Trainer.

- Give the Job Aid to each student.

At the completion of training (Lead Trainer)

- Use the evaluations and class report file (*mmddloc.doc*) to write a site report. Save the report by replacing *mmdd* in the file name with the date the class began and the *loc* with the location of the training, for example, ZDC, ZBW.

- Put the students' names next to any comments in the "Problems Encountered" section. This will allow the Volpe developers to write up a trouble ticket with a customer name and let them know Volpe is following up on their concerns.

- E-mail the site report to Joyce and Joe as soon as possible.

- Mail the original evaluations, coaching exercises, and the site report to:
  Joyce Westner
  Brattle Systems
  1100 Massachusetts Ave.
  Arlington, MA 02476

Monthly teleconference (All Trainers)

- If you are not teaching, call into the monthly teleconference.
  - First Wednesday of every month at 1 p.m. Eastern time
  - Phone: TBD

April 23, 2001

Appendix A: Preparing For TSD Training at ARTCCs

Training Sites

Site	Trainers	Address	Point of Contact & Tel.
Albuquerque ZAB	2	8000 Louisiana Blvd NE Albuquerque, NM 87109	
Anchorage ZAN	2	5400 Davis Highway Anchorage, AK 99503	
Atlanta ZTL	3	299 Woolsey Road Hampton, GA 30228	
Boston ZBW	2	35 Northeastern Blvd Nashua, NH	
Chicago ZAU	3	619 Indian Trail Rd Aurora, IL 60506	
Cleveland-Oberlin ZOB	3	326 East Lorain St Oberlin, OH 44074	
Denver ZDV	2	2211 17th Ave Longmont, CO 80501	
Fort Worth ZFW	3	13800 FAA Road Fort Worth, TX 76155	
Houston ZHU	3	6600 J F Kennedy Blvd Houston, TX 77205	
Indianapolis ZID	3	1850 S. Sigsbee St. Indianapolis, IN 46241	
Jacksonville ZJX	2	10 Aviation Ave. Hilliard, FL 32046	
Kansas City ZKC	3	250 South Rodgers Olathe, KS 66062-1689	
Los Angeles ZLA	2	2555 East Ave. Palmdale, CA 93550	
Memphis ZME	2	3229 Democrat Rd. Memphis, TN 38118	
Miami ZMA	2	7500 NW 58th St. Miami, FL 33166	
Minneapolis ZMP	2	512 Division Street Farmington, MN 55024	
New York ZNY	3	4205 Johnson Ave. Ronkonkoma, NY 11779	
Oakland ZOA	2	5125 Central Ave. Fremont, CA 94536	
Salt Lake City ZLC	2	2150 West 700 North Salt Lake City, UT 84116	
Seattle ZSE	2	3101 Auburn Way South Auburn, WA 98092	

Contacts

Name	Phone	E-Mail
Clients		
Volpe Operations Desk	(617) 494-2556	
Sylvia Harris (Training Manager)	(617) 494-2552	harriss@volpe.dot.gov
Olive Lesueur	(617) 494-2541	lesueur@volpe.dot.gov
Glenn Loop	(703) 345-8679	
Linda LaBelle	(703) 345-8676	
Volpe Center Trainers		
Justyne Johnson	(617) 494-3656	johnson@volpe.dot.gov
Christine Risko	(617) 494-2758	riskoc@volpe.dot.gov
Volpe Center Support Contractors		
Joyce Westner (Project Manager)	(781) 641-1700	JWestner@brattlesystems.com
Joe Jankowski (Team Leader)	(405) 842-1906	JJankowski@brattlesystems.com
Volpe Center Contract Trainers		
Joe Jankowski	(405) 842-1906	JJankowski@brattlesystems.com
Dag Egede-Nissen	(910) 253-1470	DEgedenisses@brattlesystems.com
Gail Griffin	(781) 641-1700	GGriffin@brattlesystems.com
Jim Champagne	(781) 641-1700	JChampagne@brattlesystems.com
Tomba Kambui	(301) 248-9131	tomba1230@yahoo.com
Charles Mohr	(405) 833-1205	CWMohr@aol.com
Sandy Prescott	(727) 360-8041	spgoldcoast@earthlink.net
Other Numbers		
Monthly Teleconference First Wednesday of every month 1 p.m. ET/EST	TBD	
TSD Hotline - Herndon, VA 7 a.m. - 7 p.m. ET/EST	703-904-4434	
TSD Hotline - Volpe 7 p.m. - 7 a.m. ET/EST	617-494-2556 617-494-2557	

April 23, 2001

Appendix B: Summary TSD Training Schedule

COMPLETED SITES

Training Begins	Site	POC & Phone & Email	Trainers & Date Traveling	No. of Students	No. of Training Terminals
9/12 – 14/2000 **PROTOTYPE**	**Washington ARTCC ZDC** 825 East Market Leesburg, VA 22075	Linda LaBelle Glenn Loop Biff Shied	**Joe Jankowski** Tomba Kambui Justyne Johnson Christine Risko Sylvia Harris Joyce Westner	29 Students - actual	3 workstations in a training area
11/1/2000 8:30 am-4 pm 3 2hr sessions 8:30 – 10:30 am 11 am – 1 pm 2 – 4 pm	**ATCSCC** 13600 EDS Drive, Suite 100 Herndon, VA 20171-3225	Bob Watkins AUA-720 (703) 326-3836	**Tomba Kambui** Justyne Johnson	9 - actual	4 workstations available in the training lab. Lab available for the entire day
11/14-16/2000 **2 Shifts** **Start** **1st shift 7:30 am** **2nd shift 1:30 pm** **There is overlap between shifts. There is flexibility with this**	**Boston ARTCC ZBW** 35 Northeastern Blvd Nashua, NH 03062	Gary O'Brien (603) 879-6468 gary_r_o'brien@faa.gov	**Chris Risko** 11/14 Joe Jankowski 11/13	24 Students - actual 3 students per session	2 PCs Control room floor

April 23, 2001

COMPLETED SITES

Training Begins	Site	POC & Phone & Email	Trainers & Date Traveling	No. of Students	No. of Training Terminals
11/27/2000 Start 1st session 8 am – 10 am 2nd session 10 am – noon One session has 2 students, the other has 3 students	ATCSCC 13600 EDS Drive Suite 100 Herndon, VA 20171-3225		Tomba Kambui (MITRE) Kelley Connelly (POC) Emily Beaton Ved Sud Elliott Simons Joe Hollenberg	5 Students - actual	6 Workstations
1/10 – 12/2001 3 instructors to conduct 3 sessions per day between 9:30 am and 5:30 pm	Salt Lake ARTCC ZLC 2150 West 700 North Salt Lake City, UT 84720	Ted Fisher Air Traffic Systems Requirements Office (801) 320-2514	Joe Jankowski Dag Egede-Nissen Tomba Kambui	49 Students - actual	3 Workstations (new PCs with 21" monitors) 1 in empty bay in control room; 2 in automation wing near control room
1/16 – 18/2001 ESIS Training – PCs w/EXCEED software Start time 7 or 8 am. Training on the day shifts	Houston ARTCC ZHU 16600 John F. Kennedy Blvd Houston, TX 77032-0032	Jerry Strickland AT Training Manager (281) 230-5510	Charles Mohr Justyne Johnson Gail Griffin	45 Students - actual	3 new PCs; training will occur on control room floor

April 23, 2001

Appendix B: Summary TSD Training Schedule

COMPLETED SITES

Training Begins	Site	POC & Phone & Email	Trainers & Date Traveling	No. of Students	No. of Training Terminals
1/23 – 25/2001 ESIS Training on all or some portion of the day and evening shifts. Training sessions: 8 - 10 am 10 am – noon 1 - 3 pm 4 – 6 pm 6 - 8 pm 8 – 10 pm if needed	New York ARTCC ZNY 4205 Johnson Avenue Ronkonkoma, NY 11779	Paul Fairley Support Manager for Training (631) 468-1053	**Dag Egede-Nissen** Gail Griffin	30 Students - actual	1 TSD workstation in TMU office
2/6 – 8/2001 3 students per session. 3 shifts per day to meet site requirements	Indianapolis ARTCC ZID Indianapolis Int'l Airport 1850 S. Sigsbee Street Indianapolis, IN 46241-3640	Chuck Whitaker Support Manager for Training (317) 247-2253	**Dag Egede-Nissen** Gail Griffin Joe Jankowski	52 Students - actual	3 PCs Training will be in the URET Lab – just off the control room floor
2/6 – 8/2001 Training sessions 10 am – 12 1 pm – 3 pm 4 pm – 6 pm	Jacksonville ARTCC ZJX 10 Aviation Avenue Jacksonville/Hilliard, FL 32046	Dan Wilmer (904) 549-1560	**Chris Risko** Charles Mohr	50 Students - actual	2 PCs (1 in an empty bay on the floor and 1 in another location)
2/13 – 15/2001	Los Angeles ARTCC ZLA 2555 East Avenue P Palmdale, CA 93550-2112	Dale Westall (661) 265-8242	**Chris Risko** Charles Mohr	38 Students - actual	2 PCs in the weather area off the control room floor
2/20 – 22/2001 3 2 hr sessions per day per instructor exact shift TBD	Atlanta ARTCC ZTL 299 Woolsey Road Hampton, GA 30228	Fred Gleason Support Manager for Training (770) 210-7814	**Dag Egede-Nissen** Justyne Johnson Charles Mohr	45 Students - actual	3 PCs Training in "Spill-out" bay of control room floor

April 23, 2001

COMPLETED SITES

Training Begins	Site	POC & Phone & Email	Trainers & Date Traveling	No. of Students	No. of Training Terminals
2/27 – 3/1/2001 Most likely 2 students per session; combination of day and evening shift 10 am – 8 pm	**Chicago ARTCC ZAU** 619 Trail Indian Road Aurora, IL 60506	Mike Pesola (630) 906-8423	**Chris Risko** Gail Griffin Charles Mohr	66 Students - actual	2 PCs off the control room floor (in the basement)
3/6 - 8/2001	**Albuquerque ARTCC ZAB** 8000 Louisiana Blvd., NE Albuquerque, NM 87109	Steve Van Sickle (505) 856-4517	**Chris Risko** Tom Alizio	26 Students - actual	2 PCs
3/13 – 15/2001	**Denver ARTCC ZDV** 2211 17th Avenue Longmont, CO 80501	Cindy Lockwood (303) 651-4113	**Gail Griffin** Charles Mohr	43 Students - actual	2 PCs off the operational floor
3/20 - 22/2001 1 instructor days 1 instructor eves	**Miami ARTCC ZMA** 7500 N. W. 58th Street Miami, FL 33166	Tom Cassady (305) 716-1535	**Joe Jankowski** Justyne Johnson	43 Students - actual	1 Workstation or PC
3/20 – 22/2001 2 Instructors (6am – 2pm) 6:30am – 8:30am 8:45am –10:45am 11:45am -1:45pm 1 Instructor (1:30pm–9:30pm) 2:00pm – 4pm 4:15pm-6:15pm 7:15pm-9:15pm	**Fort Worth ARTCC ZFW** 13800 FAA Road Fort Worth, TX 76155	Kevin Davis (817) 858-7304	**Dag Egede-Nissen** Gail Griffin Charles Mohr	67 Students - actual	4 PCs in the Emergency Operations Facility in the basement

April 23, 2001

Appendix B: Summary TSD Training Schedule

COMPLETED SITES

Training Begins	Site	POC & Phone & Email	Trainers & Date Traveling	No. of Students	No. of Training Terminals
3/27 – 29/2001 1 Instructor to work days **Desired shifts** 7:30am–9:30am 9:45am–11:45am 12:30pm– 2:30 pm 1 Instructor to work evenings **Desired shifts** 3pm–5pm 5:30pm-7:30pm 7:45pm-9:45 pm	**Memphis ARTCC ZME** 3229 Democrat Road Memphis, TN 38118	Clarence Goodwin (901) 368-8597	**Gail Griffin** Chris Risko	44 Students -actual	1 ETMS workstation in TMO's Office

April 23, 2001

TSD 2000 Coaching Exercises

STUDENT NAME(S):	
INSTRUCTOR NAME:	
FACILITY:	
DATE AND TIME:	

Part 1: Overview – Demonstrated by Trainer

Task	Accomplished
Access TSD	
Starting from the desktop, launch Exceed, log on to ETMS, and launch TSD.	
Display	
Maximize the TSD window.	
Move/Zoom to the ARTCC's airspace.	
Flights	
Show flights.	
Hide flights.	
Weather	
Show weather.	
Hide weather.	
Alerts	
Show alerts.	
Hide alerts.	
Reroute	
Show reroutes.	
Hide reroutes.	
Exit	
Close TSD, ETMS, and Exceed.	

Part 2: Windows Basics – Performed by Trainee

Task	Accomplished
Use Windows Basics	
Launch Exceed, log on to ETMS, and launch TSD.	
Mouse buttons.	
Single and double clicking.	
Active windows.	
Moving windows.	
Resizing windows.	
Maximizing and minimizing windows.	
Closing windows.	
Menu bar and mnemonics.	
Tear off menus.	
OK, Apply, and Cancel buttons in dialog boxes.	

Part 3: TSD

Task	Accomplished
Use the Maps Menu	
Adjust the map so the ARTCC airspace fills the display.	
Zoom out to see appropriate handoff area.	
Display desired overlays and labels (e.g., pacing airports, sectors, fixes).	
Display one or more map items.	
Save your map items in a folder.	
Display range rings around a fix.	
Save your range ring settings in a folder.	

April 23, 2001

Task	Accomplished
Use the Display Menu	
Show the Times window.	
Add a legend.	
Change the color of at least one overlay.	
Change the font size of at least one label.	
Save your display using the Adapt command. Record the name here:	
Recall the default display.	
Recall your display.	
Use the Flights Menu	
Show flights.	
Display arrivals and departures, in different colors, for an airport in your area. Enable Show Route only.	
Hide all other flights.	
Display the Flight Count window.	
Select the automatic icon for the arrivals.	
Copy your arrivals flight set. Change the color and add an appropriate departure airport.	
Delete all flight sets except the arrivals.	
Add a new flight set. Display all heavies transiting your ARTCC or sector in a new color.	
Change the priority of the flight sets using the Top command and Click-and-Drag.	
Save your flight sets in a folder.	
Use the Browse Flights feature to display temporarily the data block for a flight transiting your ARTCC or sector. Make a note of the ACID.	
Hide flights.	
Use the Find Flight command to display, in yellow, the aircraft you noted earlier.	
Find all American Airlines flights using the Wildcard function.	

Task	Accomplished
Remove all found flights.	
Show flights.	
Save your display using the same name as before.	
Use the Flight Icon Pop-up Menu	
Display a data block for a flight.	
Toggle between Org/Dest and Route in the data block.	
Draw the flight plan route.	
Activate history tracking for a flight.	
Display the Last TZ for a different flight.	
Delete an aircraft icon.	
Change the color of a flight with a displayed data block.	
Reposition the data block.	
Examine the flight with the history track.	
Hide flights.	
Use the Alerts Menu	
Show alerts.	
Display the alerts for your type of sector.	
Examine an alerted sector.	
Display the alerted sector's bar chart.	
Display the alerted sector's alerts report.	
Display the flights for the alerted sector.	
Close all alerts windows.	
Hide alerts.	
Use the Weather Menu	
Show weather.	
Select NOWRAD 2km for your area. Include Canadian or San Juan weather, if appropriate. Verify NOWRAD Legend is on and Lightning, Tops, and Jet Stream are off.	

Task	Accomplished
Show only precipitation level 3 and above.	
Display radar tops in blue.	
Display the latest lightning in white.	
Display all three jet stream options.	
Hide weather.	
Use the Reroute Menu	
Show reroutes.	
Create and name a private reroute.	
Turn off the reroute just created.	
Delete the reroute just created.	
Hide reroutes.	
Use the Tools Menu	
Show flights.	
Print the display.	
Request a Count.	
Request a List.	
Request an Area.	
Request an ARRD.	
View Multiple TSDs	
Launch a second TSD window.	
Arrange the TSD windows so the projected display is in the lower right corner.	
Exit	
Close TSD, ETMS, and Exceed.	

Part 4: Setting up Your Display

Task	Accomplished
Create an Individual Display	
Launch TSD.	
Recall your saved display.	
Display desired overlays.	
Display your own map items and save.	
Create your own range ring settings and save.	
Create your own flight sets and save.	
Display desired weather settings.	
Display desired alerts.	
Create and display desired reroutes.	
Launch a second TSD window and arrange the desktop.	
Save your display.	
Close TSD, ETMS, and Exceed.	

Notes

Use the space below to record problems encountered during training. Be specific.

TSD 2000 Coaching Exercises
Instructor Version

Part 1: Overview – Demonstrated by Trainer

Task
Access TSD
Starting from the desktop, launch Exceed, log on to ETMS, and launch TSD.
• By default, the TSD window should not be maximized and should display a Black Background with Gray Boundaries and Brick Red ARTCCs (labeled).
Display
Maximize the TSD window.
Move/Zoom to the ARTCC's airspace.
• Place cursor in the center of the ARTCC's airspace and press **Z**.
Flights
Show flights.
• By default, no flight sets are selected. Point out that flights are displayed only if some flights are selected. Open the Select Flights window, click the Show box for ATL arrivals, then click OK.
Hide flights.
Weather
Show weather.
• Point out that weather is displayed only if weather is selected.
Hide weather.
Alerts
Show alerts.
• Point out that alerts are displayed only if alerts are selected.
Hide alerts.
Reroute
Show reroutes.
• Point out that reroutes are displayed only if reroutes have been created.
Hide reroutes.

Appendix D: TSD 2000 Coaching Exercises – Instructor Version

Task
Exit
Close TSD, ETMS, and Exceed.

Part 2: Windows Basics – Performed by Trainee

Task
Use Windows Basics
Launch Exceed, log on to ETMS, and launch TSD.
Mouse buttons. - Explain the left (primary) button is used for routine clicking, just as in Windows. - The right button is used to display pop-up windows, just as in Windows. - The middle button is for moving items, such as data blocks. - On PCs with only a two-button mouse, use the right button for moving items.

Single and double clicking.
- Point out that most actions require only a single click.
- Some double click actions are:
  - Display a data block (double click a flight icon)
  - Highlight text in a dialog box (will show in the Move/Zoom window).
  - Close a window (double click the Window Menu button on the far left side of the Title Bar). Close the Times window using this method.
- Point out the Other Functions section of the Job Aid.

Active windows. - Point out the TSD window is active (TSD Title Bar in color, Times Title Bar grayed out). Make the Times window active.
Moving windows. - Move the Times window.
Resizing windows. - Resize the TSD window by click-dragging a side, bottom, and corner. - Point out after the TSD window is resized, the display re-centers.
Maximizing and minimizing windows. - Maximize the TSD window, then minimize, then maximize again.
Closing windows. - Close the Times window by selecting Close after clicking the Window Menu button.

Task
Menu bar and mnemonics. • Click on the Display menu. Point out the choices. • Point out the ... signifies that a dialog box will be displayed if selected. • Point out the hollow arrowhead signifies that another menu will be displayed if selected. • Close the Display menu. Point out the underlined **D** in the word Display. • Press **Alt D** to display the menu. Press **T** to open the Times window.
Tear off menus. • Tear off the Display menu by clicking on the dashed line and move the window. • Close the Tear-Off window by right-clicking the Title Bar and choosing Close.
OK, Apply, and Cancel buttons in dialog boxes. • Display the Select Flights window. • Explain what happens when each button is clicked.

Part 3: TSD

Task
Use the Maps Menu
Adjust the map so the ARTCC airspace fills the display. • Open the Move/Zoom window. • Place the cursor on the center of the ARTCC airspace and type **M**. • Point out the lat/lon of the cursor point is displayed in the Location box. Tell the trainees about the other centering options. • Single click the Zoom Scale box to activate. • Double click the numbers to highlight. • Type new zoom scale numbers (just estimate what is needed). • Click the Apply button. • Repeat until the airspace complete fills the display. • Close the Move/Zoom window.
Zoom out to see appropriate handoff area. • Repeat previous steps but increase the number by 100 miles. • Ask if enough buffer is now displayed. • If not enough, increase the scale until the trainee is satisfied with the buffer.

Task
Display desired overlays and labels (e.g., pacing airports, sectors, fixes). <ul><li>Open the Map Overlay window.</li><li>Ask the trainee what overlays he/she wants displayed.</li><li>You want a set of sectors, fixes, or navaids displayed at least temporarily in order to show how cluttered the display can get.</li></ul>
Display one or more map items. <ul><li>Encourage the use of map items to reduce screen clutter.</li><li>Ask the trainee to display the individual sectors, fixes, and/or navaids he/she uses. Insure at least one fix is displayed.</li><li>After individual map items are displayed, turn off the matching overlays.</li></ul>
Save your map items in a folder. <ul><li>Save as **Mapitems**/Filename, where Filename is the trainee's last name (capitalize the first letter).</li><li>Point out that the list of folders/files is sorted in ASCII order (caps appear before lowercase).</li></ul>
Display range rings around a fix. <ul><li>If the trainee has a preference, use it.</li><li>Otherwise, show two labeled, light gray, 10 mile rings around one fix.</li><li>Point out that range rings can be used to display a Temporary Flight Restriction.</li></ul>
Save your range ring settings in a folder. <ul><li>Save as **Rangerings**/Filename, where Filename is the trainee's last name (capitalize the first letter).</li></ul>
Use the Display Menu
Show the Times window. <ul><li>Explain that the window will show the update times for flights, alerts, weather, and reroutes.</li><li>Move the Times window to the lower right corner, but leave space for it to expand.</li></ul>
Add a legend. <ul><li>Use the trainee's last name, but explain that it can contain anything.</li></ul>
Change the color of at least one overlay.
Change the font size of at least one label.
Save your display using the Adapt command. Record the name here: <ul><li>Save as **Displays**/Filename, where Filename is the trainee's last name (capitalize the first letter).</li></ul>
Recall the default display.

Task
Recall your display. • Point out that the procedure for recalling saved map items and range rings is similar—invoke the recall command under the File menu on the Show Map Items and Select Range Ring windows.
Use the Flights Menu
Show flights. • Point out the update time in the Times window.
Display arrivals and departures, in different colors, for an airport in your area. Enable Show Route only. • You may want to move the window to one corner and click Apply.
Hide all other flights. • Click the checked Show boxes.
Display the Flight Count window. • Explain it.
Select the automatic icon for the arrivals. • Explain the icons.
Copy your arrivals flight set. Change the color and add an appropriate departure airport. • Point out that both sets of arrivals are displayed in different colors (zoom out if necessary). • Explain that this occurs because the subset of all arrivals (the one with a departure airport) has a higher priority.
Delete all flight sets except the arrivals.
Add a new flight set. Display all heavies transiting your ARTCC or sector in a new color. • Point out that some previously displayed flights may change color. • Point out that this could be useful when weather causes flights to be rerouted through the trainees sector.
Change the priority of the two sets of arrivals using the Top command and Click-and-Drag. • Top line, arrivals with a departure airport. • Second line, all arrivals. • Third line, transiting heavies.
Save your flight sets in a folder. • Save as **Flights**/Filename, where Filename is the trainee's last name (capitalize the first letter). • To recall a saved flight set, invoke the Recall command from the File menu.

Task
Use the Browse Flights feature to display temporarily the data block for a flight transiting your ARTCC or sector. Make a note of the ACID.
Hide flights.
Use the Find Flight command to display, in yellow, the aircraft you noted earlier. • Point out that by using yellow, the flight really stands out.
Find all American Airlines flights using the Wildcard function. • Point out the differences between * and ?.
Remove all found flights.
Show flights.
Save your display using the same name as before.
Use the Flight Icon Pop-up Menu
Display a data block for a flight. • Double click a flight icon
Toggle between Org/Dest and Route in the data block.
Draw the flight plan route.
Activate history tracking for a flight. • Point out that the icon must move before history is displayed.
Display the Last TZ for a different flight. • Stress that the TZ point is the last radar hit; since then, the flight's position has been extrapolated based on the flight plan.
Delete an aircraft icon. • Redisplay the icon by hiding flights, then showing flights again.
Change the color of a flight with a displayed data block.
Reposition the data block.
Examine the flight with the history track. • Zoom in if necessary.
Hide flights.
Use the Alerts Menu
Show alerts.
Display the alerts for your type of sector. • Check the desired sector type and uncheck Airports All.

Task

Examine an alerted sector.
• Explain how to distinguish between low (\\\\), high (///), and superhigh (|||).
• Explain the different colors:
    ▪ Red – The flight that caused the alert is airborne.
    ▪ Yellow – The flight that caused the alert is not yet airborne.
    ▪ Green – The controllers for that sector, fix, or airport told the TMU that they could handle the flights.
    ▪ Purple – The sector, fix, or airport is not alerted but is being examined.

Display the alerted sector's bar chart. • Point out that red denotes the number of airborne flights and yellow the number still on the ground. • The capacity level is set by the controlling agency and input by the TMU.
Display the alerted sector's alerts report. • Point out that this is a list of flights transiting the sector during each 15-minute time period specified.
Display the flights for the alerted sector. • Point out that this displays the flights transiting the sector during each 15-minute period specified. • Point out the circles at airports where aircraft are still on the ground. • Explain the numbers (number on the ground/number past their original take-off time and under a ground delay program/number on the ground past their ETD). • Highlight a circle to display the ACID of the flights on the ground and the proposed or scheduled take-off times.
Close all alerts windows.
Hide alerts.
Use the Weather Menu
Show weather.
Select NOWRAD 2km for your area. Include Canadian or San Juan weather, if appropriate. Verify NOWRAD Legend is on and Lightning, Tops, and Jet Stream are off. • Zoom out if necessary to see weather. • Weather comes from the National Weather Service through the Command Center.
Show only precipitation level 3 and above.
Display radar tops in blue.   • Point out that this allows you to distinguish between rain showers and thunderstorms.

Task
Display the latest lightning in white.
• White allows you to see the lightning within weather.
Display all three jet stream options.
• Explain the display.
Hide weather.
Use the Reroute Menu
Show reroutes.
Create and name a private reroute.
Turn off the reroute just created.
• Uncheck the Show box in the Select Reroutes window.
Delete the reroute just created.
• Use the Edit menu in the Select Reroutes window.
Hide reroutes.
Use the Tools Menu
Show flights.
Print the display.
• Change the background color to white in order to conserve ink.
• You may need to darken other colors for contrast.
• Stress that you have to Capture if you want to print the current display.
• Recall a saved display.
Request a Count.
• Type **REQ** sectorname or artccname, e.g. **REQ ZDC12**.
Request a List.
• Type **REQ** sectorname or artccname **LIST**, e.g. **REQ ZDC12 LIST**.
Request an Area.
• Type **REQ AREA** [sectorname sectorname], e.g. **REQ AREA [ZDC12 ZDC58]**.
Request an ARRD.
• Type **REQ ARRD** icao code, e.g. **REQ ARRD IAD**.
View Multiple TSDs
Launch a second TSD window.

Task
Arrange the TSD windows so the projected display is in the lower right corner. • Point out that by projecting the display in the lower right corner, the supervisor can use the other TSD window to request reports, display alert timelines, etc., without interfering with the projected display.
Exit
Close TSD, ETMS, and Exceed.

Part 4: Setting up Your Display

Task
Create an Individual Display
• Allow the trainee to work on his/her own, but answer any questions the trainee may have.
Launch TSD.
Recall your saved display.
Display desired overlays.
Display your own map items and save.
Create your own range ring settings and save.
Create your own flight sets and save.
Display desired weather settings.
Display desired alerts.
Create and display desired reroutes.
Launch a second TSD window and arrange the desktop.
Save your display.
Close TSD, ETMS, and Exceed.

Appendix E: TSD for Controllers Coaching Exercise Evaluation

TSD for Controllers Coaching Exercise Evaluation

NAME AND POSITION:	
INSTRUCTOR NAME:	
FACILITY:	
DATE AND TIME:	

Please circle the appropriate number for each question below:

1. My ability to use TSD has increased.

1	2	3	4	5
Strongly Disagree	Disagree	Neutral	Agree	Strongly Agree

2. The instructor presented information clearly and understandably.

1	2	3	4	5
Strongly Disagree	Disagree	Neutral	Agree	Strongly Agree

3. The instructor appeared knowledgeable about the TSD software.

1	2	3	4	5
Strongly Disagree	Disagree	Neutral	Agree	Strongly Agree

4. This training will help me in my job.

1	2	3	4	5
Strongly Disagree	Disagree	Neutral	Agree	Strongly Agree

5. What was the most valuable part of this training?

   _____

   _____

6. What additional information would you like to see covered in future training?

   _____

   _____

7. How could this training be improved?

   _____

   _____

8. Please add any other comments you would like to make. (For example, spend more or less time on certain topics; add or eliminate certain topics; speed up or slow down the pace of the training, etc.)

   _____

   _____

April 23, 2001

TSD Familiarization Prototype Evaluation Report

Dates	September 12, 2000
Location	ZDC
Population	3 students
Trainer	Joe Jankowski
Observers	Sylvia Harris, Linda LaBelle, Glenn Loop, Tomba Kambui, Justyne Johnson, Christine Risko, Joyce Westner, Biff Shied

Students

Mark McFarland (CIC) Randy Artis (SATCS)
Phillip S. Swinney (MOS)

Special Factors

The training was conducted with three students sharing one workstation in order to prototype the worst scenario that could occur in the field.

Evaluation Feedback

The following table lists the students' evaluation of the class, based on the responses of the 3 students.

Question	Average Response
1. My ability to use TSD has increased	4.67
2. Instructors clear and understandable	5.00
3. Instructors knowledgeable	5.00
4. Training will help in my job	4.67

Scores are on a scale of 1-5: where
1 = Strongly Disagree 2 = Disagree 3 = Neutral 4 = Agree 5 = Strongly Agree

Continued on next page

TSD Familiarization Prototype Evaluation Report, Continued

Most valuable part of course	The students responded: • Learning how to adapt each TSD to new traffic situations • Hands on/actually using the equipment • Personalizing my TSD display
What to include in future training?	The students responded: • N/A • N/A • More hands on time
How could training be improved?	The students responded: • Make this training mandatory • N/A • More exercises.
Other comments	The students responded: • N/A • N/A • Overall very good
Suggestions on topics to remove	The prototype students recommended that we keep most of the tasks, but remove the following: • Email • Finding the distance between two points and to a third point using DME • One student suggested removing the tear off menu, the range rings, and the flight route, but other students and observers recommended keeping them.

Continued on next page

TSD Familiarization Prototype Evaluation Report, Continued

Suggestions on topics to change

The prototype students recommended that we keep most of the tasks, but change the following:
- Center the map during the first demonstration of Move/Zoom; move to the VORTAC later
- Explain that range rings may be useful during a TFR temporary flight restriction
- Add an activity so students create, show and hide a reroute. When teaching reroutes; show which reroutes might affect your flow, e.g., Pittsburgh flights may be rerouted through a ZDC sector. Or Jet 48 in sectors 4 and 72 may be impacted by Atlanta arrivals. Or the weather may push flights into sector 3.
- Emphasize differences between mouse buttons [also point it out on the Job Aid]
- Describe the three types of flight icons [perhaps on the Job Aid]
- Show arrivals to a center or enroute through a sector
- Explain the 2 wildcards in more depth
- Explain "last TZ" for a flight
- When examining the flight with the history track, also examine the flight with the "last TZ"
- Explain what the alerts colors mean [and add explanation to the Job Aid]; also explain that local TMU sets the capacity level
- Explain what to look for in the bar chart
- Explain the difference between NOWRAD 2km and 8km weather
- Demonstrate multiple TSD windows
- Explain how the projector may show only one window or one part of a window
- Show how a report can be displayed without disrupting the window that's being projected on the wall; e.g., use the projector to display only the lower right quarter of the TSD screen.

TSD Familiarization Evaluation Report - ZDC

Dates	September 13-14, 2000
Location	DC ARTCC
Population	26 students
Trainers	Tomba Kambui, Justyne Johnson, Christine Risko, Joe Jankowski

Students

Steve Privott	Richard Ossana
L'Tanya Talley	Gary Bukovskey
David Steele	Bob Staudenmeirer
Charlyn Davis	John Brame
L. Sue Walden	George Greenfield
Roger Gilmore	Mike Goldser
Gregg White	Biff Schied
Chris Sutherland	Jeffrey Wellborn
Jackie Stell	Jeff Weldon
Al Anch	Bob Bearer
Drew Kassal	Bill Pearman
Deborah Archut	Michael DeMonte
Thomas Reiss	Kathy Brommage

Special Factors — Trainers were available for 12 blocks of coaching each day-- two trainers at 8, 10, and 12; two at 2, 4, and 6. This allowed the trainers to coach every available supervisor and CIC.

Continued on next page

TSD Familiarization Evaluation Report - ZDC, Continued

Evaluation Feedback

The following table lists the students' evaluation of the class, based on the responses of the 20 students.

Question	Average Response
1. My ability to use TSD has increased	4.69
2. Instructors clear and understandable	4.81
3. Instructors knowledgeable	4.88
4. Training will help in my job	4.62

Scores are on a scale of 1-5: where
1 = Strongly Disagree 2 = Disagree 3 = Neutral 4 = Agree 5 = Strongly Agree

Most valuable part of course

The students responded:
- Being able to set up the display myself.
- How to select desired displays.
- Learning how to save my customized screen
- hands-on training
- The entire session was very educational.
- Hands on with competent instructor
- Select Flights & Alerts
- Increased familiarity with how to use the software.
- Learning to monitor different traffic flows
- Since I didn't know anything, this was great.
- Hands on learning
- Hands on training
- Learning how to utilize data that has been available in my area and manage resources better.
- Hands on personalized training
- Hands on experience
- Learning to display what I need when I want it
- How to modify flight selection and save display settings
- Alert ability and weather display capability
- Learning different aspects of TSD
- Learning different functions to help me on the floor.

Continued on next page

TSD Familiarization Evaluation Report - ZDC, Continued

Most valuable part of course, continued

The students responded:
- Hands on
- Request command line
- The total hands-on approach, being able to (play) with the system enhances the learning curve.
- hands on training
- To become familiar with the TSD and various commands

What to include in future training?

The students responded:
- more filter info
- combine this training with ESIS
- Include in depth ESIS training
- Cheat Sheets
- Expand info to include specific "ATC Traffic Problems" and how to use TSD to solve/address the "problem".
- How to put it all together and show the entire picture.
- I would like to have this training again in a few months.
- how the information will be input and accessed from the area.
- More often training
- Maybe instruct toward Area Specifics (Example Area 8 NY area ideas etc...)
- Arrival and departure lists

Continued on Next Page

TSD Familiarization Evaluation Report - ZDC, Continued

How could training be improved?

The students responded:
- It's very good now.
- Good training.
- I believe it could be a little longer
- Include ESIS training
- CBI precede individual training
- Use of "Cheat Sheets" or reference cards that you could take with you for review
- The training was adequate. No improvement necessary.
- Training could be done quarterly.
- TMU could explain their input and how they would use it.
- No improvement necessary
- Less rushed more time
- Additional time with equip
- I can't think of any, very nice job!
- It was great – very informative.

Other comments

The students responded:
- Good pace. Suggest that training (refresher) be made available in future.
- No recommendations, instruction was very helpful.
- Instructors were knowledgeable and courteous.
- Include ESIS training. Provide training in more timely manner
- Excellent course!
- Pace & material seem @ appropriate level.
- It would be nice if the class was longer to afford slower people more hands on time.
- Nice Job
- well done enough information to use the TSD
- Slow pace
- Thank you!!
- Only had CBI lesson on system – time spent was adequate for experience
- Everything was good. Nice job "Tomba"

TSD Familiarization Evaluation Report - ATCSCC

Dates	November 1, 2000
Location	Command Center
Population	9 students
Trainers	Tomba Kambui and Justyne Johnson

Students	Bob Watkins Dick Sullivan James G. Garlitz George Curley Ernie Leonard Jim Diehl Teri Bristol Geoff Shearer Jim Houde

Special Factors	Trainers were available for 6 blocks of coaching -- two trainers at 8:30, 11, and 2. When more than one student attended a session with an instructor, they shared a workstation and took turns practicing activities in TSD.

Continued on next page

TSD Familiarization Evaluation Report - ATCSCC, Continued

Evaluation Feedback

The following table lists the students' evaluation of the class, based on the responses of the 9 students.

Question	Average Response
1. My ability to use TSD has increased	4.66
2. Instructors clear and understandable	4.88
3. Instructors knowledgeable	4.66
4. Training will help in my job	4.44

Scores are on a scale of 1-5:   where
1 = Strongly Disagree   2 = Disagree   3 = Neutral   4 = Agree   5 = Strongly Agree

Most valuable part of course

The students responded:
- Overall, everything was valuable since I didn't have familiarity with TSD but the alerts and reroutes were <u>most</u> valuable.
- General familiarity.
- The in-depth explanation of capabilities and functionality.
- Hands-on/person to person.
- Hands on experience with features in TSD.
- Hands on.
- Overall general knowledge.
- Ability to observe individual tracks.
- Hands on.

Continued on next page

TSD Familiarization Evaluation Report - ATCSCC, Continued

What to include in future training?	The students responded: • Nothing I can think of right now. • Other features besides TSD. • OK as is. • Additional ETMS products, e.g., FSM. • Written explanation to take. • Would like to be able to see departure times for individual aircraft.
How could training be improved?	The students responded: • Needs to be recurring since we don't work with the equipment. • No recommended improvements. • No comment. • No suggestions. • Individual displays for each student.
Other comments	The students responded: • It was fine. Thanks! • Excellent instruction. • About right! • Good instructor!

TSD Familiarization Evaluation Report - ZBW

Dates	November 14-16, 2000
Location	Boston ARTCC
Population	24 students
Trainers	Christine Risko and Joe Jankowski

Students

Frank Dorman
Charles Lavigne
Patricia Singletary
Don Stetz
Layne LaBaume
Ron McEwen
Ray Newman
Danny L. Ireland
Tim Sowder
Jack Aiken
Dave Emmes
Randy Wood

Tim Powers
Stephen Marlar
Mark McKelligan
Greg Bull
J. Heinz
Ron Smith
Chris Allison
Jon Schippani
M. Lyman
John Fitzgerald
Keith McKinley
Maxine Rosa

Special Factors Trainers were available for 6 blocks of coaching each day—8 and 10 a.m., and 1, 2, 4, and 7 p.m. A regularly scheduled monthly event, ACTeam training, was conducted Wednesday. This reduced the availability of trainees. The cubicles in which training was conducted were too small for a coach and three students. In addition, the ETMS equipment was experiencing technical problems so the monitors used for training were limited to 800x600 resolution. This caused many of the dialog boxes to be incorrectly displayed.

Evaluation Feedback The following table lists the students' evaluation of the class, based on the responses of the 20 students.

Question	Average Response
1. My ability to use TSD has increased	4.79
2. Instructors clear and understandable	4.96
3. Instructors knowledgeable	4.83
4. Training will help in my job	4.54

Scores are on a scale of 1-5: where
1 = Strongly Disagree 2 = Disagree 3 = Neutral 4 = Agree 5 = Strongly Agree

Continued on next page

TSD Familiarization Evaluation Report - ZBW, Continued

Most valuable part of course

The students responded:
- Just the overall lesson of TSD was valuable and interesting.
- Good overview of alert and list functions
- Hands-on
- Reports
- The ability to see further outside the center's airspace to forecast traffic.
- Learning how it operates
- Hands on time
- Very knowledgeable instructor. One on one (two) situation instead of large class allowing hands-on training.
- On hands
- Introduction to TSD functions and their use
- Learning routes, types and wx [weather]
- Re-learn what I have not used
- Hands on
- Hands on
- The alerts and weather
- Actual use of equipment
- Hands on participation.
- Learning techniques to predict sector workload.
- I have had very little time with TSD to this point. So anything is better than before
- Hands on
- Coming to realize how much information is available and how it can make my job easier.
- Ability to change displays, understand commands
- The TSD Menu Functions handout
- Hands on with well paced coaching – not too fast

Continued on next page

TSD Familiarization Evaluation Report - ZBW, Continued

What to include in future training?	The students responded: • The effects of reroutes on A/C through TSD • Just right. • None • Probably none – it seemed like we covered all important info, and extra info • None – I may find some as I use the system. • More time and detail • A little more time for hands on • Everything was really good. • None • All information was covered adequately; instruction on how to integrate it into the job should be the next step.
How could training be improved?	The students responded: • Possibly (although I hate to say this) a CBI lesson to get more accustomed to all the features • Training was fine • The training seems fine as is. • Training was fine • Incorporate a short break – I was starting to zone out a little the last 10 min. • Using the computer – on hands for each student • CBI refresher • Just right • Hand outs of items covered for memory joggers [The Job Aid was given to all students] • This is a nice introduction to TSD – compact yet informative. After the users have some experience with this equipment, a second, more advanced, training session would be helpful. • Allow time for experimentation. • More practice would help • All information was covered adequately; instruction on how to integrate it into the job should be the next step.

Continued on Next Page

TSD Familiarization Evaluation Report - ZBW, Continued

Other comments

The students responded:
- Great job
- Training was fine
- Nice Job
- Good Job Christine!
- Great pace considering we are familiar with computers
- I thought the course was better presented than any software programs that I have seen before
- Pace and topics were just about as good as they could be. Coordinate w/supcom (Rick Winch) on how to integrate knowledge into job skills.
- V. G. Job.
- Professionally presented – Thank you

TSD Familiarization Evaluation Report - ATCSCC

Dates	November 27, 2000
Location	ATCSCC
Population	5 students
Trainers	Tomba Kambui

Students	Elliott Simons Kelly Connolly Emily Beaton Joe Hollenberg Ved Sud

Special Factors	The trainer was available for 2 blocks of coaching—8 and 10 a.m. Only one terminal was available.

Evaluation Feedback	The following table lists the students' evaluation of the class, based on the responses of the 5 students.

Question	Average Response
1. My ability to use TSD has increased	4.80
2. Instructors clear and understandable	4.80
3. Instructors knowledgeable	4.40
4. Training will help in my job	4.60

Scores are on a scale of 1-5: where
1 = Strongly Disagree   2 = Disagree   3 = Neutral   4 = Agree   5 = Strongly Agree

Most valuable part of course	The students responded: • Flights and alerts • Hands-on training • Reports feature and overview of all functions • Flights/alerts; hands on is good • Primary introduction

Continued on next page

TSD Familiarization Evaluation Report - ATCSCC, Continued

What to include in future training?	The students responded: • Other ETMS functions • TSD rerouting capabilities • Reroutes • Rerouting and reports • N/A
How could training be improved?	The students responded: • Would like cheat sheet up front to take notes if wanted • Need to think about it
Other comments	The students responded: • Thanks a lot • Instructor was good—patient and humorous • Good Job

TSD Familiarization Evaluation Report - ZLC

Dates	January 10-12, 2001
Location	ZLC- Salt Lake ARTCC
Population	49 students
Trainers	Joe Jankowski, Tomba Kambui, Dag Egede-Nissen

Students

Jim Hopkin (OS)	Dave Adams (ATCS)
Valerie Nolan-Stuhr	Steve Pearce
Dayna Parks (CPC)	Gordon Flygare (Tech)
Andy Andrelezyk (ATCS)	Rocco Vitacca
Michael R. Jankorich (Sys Spec)	Steven Christiansen (ATCS)
Will Andrews (OM)	Daniel M. Cook
Chris Peterson (ATC)	Carl Knapp (ATCS CIC)
Andy Flores (SS)	Dave Skidmore (Tng Spec)
Paul Vawdrey (Tng Spec)	Richard Seaman (ATCS)
R Yerman	Doug Barrow (ATCS)
Bob Eck	Charlie Palmer
Keith R. Buys	Scott Jorgensen (ZLC533)
Dave Molyneux (ATCS)	Mike Blackburn (ATCS)
Steve Olsen (ATCS)	Doug Davis (Sys Spec)
Jim Burrows (ATCS CIC)	Jill Story
Rodney Merrill (ATCS)	Chris Winterroth
Milt Rassmussen	Greg Times
Art Myshrall	Kenneth B Tackett
Deborah Bradford	Kevin Thomas (ATCS CIC)
Robert Elliott (ATCS CIC)	Wayne E Demke (Lock-Mart FME)
Blaine Hill (ATCS)	Cory Olsen (ATSS)
Steve Phillips (ATCS)	Joel Stockseth (Supervisor)
Minh Q. Nguyen	Troy Decker (ATCS)
Mark Scorel (ATCS)	Grant Johnson (ATCS)
Tracy Casil (ATCS)	

Continued on next page

Appendix F: TSD Familiarization Evaluation Reports

TSD Familiarization Evaluation Report - ZLC, Continued

Special Factors

The training was conducted at three stations, one in a bay at the end of the work bays and two in a separate room.

Computers were fitted with the ESIS system, which allows a display to be divided into two parts, half on one computer screen and half on another screen. Trainees were given a 30-minute training session on this system immediately prior to receiving the TSD training. The ESIS system facilitates the splitting of displays between computer and projector screens, as well as the use of multiple TSD displays on two computer screens simultaneously. Nearly all TSD training was conducted on a single screen, although at the end of each session, trainers showed students how to drag a display from one computer screen to the other.

The ZLC site coordinators (Ted Fisher and Paul Vawdrey) had done an impressive job in establishing a training schedule that was adhered to in a timely and effective manner. Student experience with TSD varied from having worked with either TSD or ASD in TMUs, to having completed the CBI, to having no experience (approximately half). All students displayed a positive approach to the training.

Evaluation Feedback

The following table lists the students' evaluation of the class, based on the responses of the 3 students.

Question	Average Response
1. My ability to use TSD has increased	4.57
2. Instructors clear and understandable	4.78
3. Instructors knowledgeable	4.76
4. Training will help in my job	4.41

Scores are on a scale of 1-5:   where
1 = Strongly Disagree   2 = Disagree   3 = Neutral   4 = Agree   5 = Strongly Agree

Continued on next page

April 23, 2001

TSD Familiarization Evaluation Report - ZLC, Continued

Most valuable part of course

The students responded:
- Hands on practice with the equipment
- Hands on use of the equipment
- Hands on training
- Hands on experience
- Demo – show & tell; Hands on
- That I was invited to sit in on this training. It will help when I am hired.
- Hands on.
- Hands on training
- Hands-on
- Hands on use of computer
- Good interaction with instructor
- 1 on 1 hands on
- Hands on training
- Hands on – being able to use the equipment
- Review. I had previously worked with TSD
- Just gave a person who had no knowledge of TSD some new light
- Hands on practice
- Familiarization with software capabilities
- Hands on
- Hands on instruction
- Update of the software
- Hands on
- The hands on training and being able to execute the commands
- Hands on time
- Hands on work
- Adjusting/modifying & saving of individual script
- Provided a good introduction to TSD
- One on one
- 1 vs 3 ratio – and hands on training
- Hands on
- One on one training
- TSD familiarization
- All
- Hands on OJT with instructor
- Personalized training with only two students at a time. Joe was very helpful and knowledgeable.

Continued on next page

TSD Familiarization Evaluation Report - ZLC, Continued

Most valuable part of course (continued)

The students responded:
- Familiarization with shortcut commands
- The hands on training
- Hands on training
- I had almost zero knowledge to start out. Excellent overview of functions
- "Hands-on" experience
- All
- Hands on time and learning capabilities
- Hands on training
- Being able to access information on a real time basis – (learning how to).

What to include in future training?

The students responded:
- More facility specific
- Don't know until I start using the system
- N/A
- Short-cut commands and cheat sheet
- Just the hands on is sufficient
- Maybe a training guide
- Hot keys/short cuts
- None
- None
- For what I will be doing, the training was complete.
- The training seemed like it was geared to TMU. Could use a little more emphasis on ATCS functions.
- Nothing.
- More time to play/practice. Individual PC for each student.
- None
- Best way to train personnel
- System architecture
- Maybe just review
- None
- The information presented was TMU directed. It would be more useful to me if it was more Area/Sector specific.

Continued on next page

TSD Familiarization Evaluation Report - ZLC, Continued

How could training be improved?

The students responded:
- Provide a few breaks to help knowledge overload.
- Trainer did a great job
- Don't know
- Seemed very good
- N/A
- More hands on time
- N/A
- Dinner and movie afterwards
- Possibly a stronger emphasis on ATC orientation. Maybe a trained controller available also.
- I like one on one training. It is the best way to teach most people.
- More time to play with the unit on my own.
- Fix the bugs in the program prior to the training
- If all the functions would have worked
- Reduce class size to 1:1 or provide each controller w/ TSD to walk through w/ instructor
- More hands on time
- Other than holding it in Cambridge, MA, nothing
- More time
- Can't think of any way
- I don't see any necessary improvement. I liked the course very well.
- None
- It was too short
- Even more "hands on" time
- Have bugs out of the system before training begins

Continued on next page

TSD Familiarization Evaluation Report - ZLC, Continued

Other comments

The students responded:
- Face to face instruction was very helpful
- Liked the one-on-one format. Pace was just right.
- I like one on one or small hands on training. Hands on training helps me retain info better.
- I like the low student;instructor ratio
- Good course
- Good training
- One on one training worked well in this situation
- Pace was perfect
- Good job.
- Possibly a little more hands on time
- Small group helped, made it easier to ask questions. Should be a longer session.
- Keep class size small w/ hands on experience.
- Pace was a little fast – had to hurry last few items
- Need low numbers – one on one.
- Instructor was very knowledgeable. I especially liked the one on one training.
- It was just perfect.
- Very happy with its current pace.
- Follow up training to insure information obtained today is not forgotten from too much time between training and application.

Appendix F: TSD Familiarization Evaluation Reports

TSD Familiarization Evaluation Report - ZHU

Dates	January 16-18, 2001
Location	ZHU - Houston ARTCC
Population	45 students
Trainers	Justyne Johnson, Charles Mohr, and Gail Griffin

Students

- Keith Bradley
- George E. Dick (OS)
- Rocky Lane (OS)
- Rande Hoffman (CPC)
- Bernadette Session (OS)
- Bill Smayda
- Alton Slayton (CRP - OS)
- Michael Kingsley
- Ken Wilson (RSG - OS)
- Richard Rance
- Carlos Gonzalez (Supervisor)
- Larry Wanless
- Ron Pyron
- Alex Herda (ATCS)
- Leslie Echols (CPC)
- Steve Stooksberry (OS)
- John Boutté
- S. Price (Supervisor)
- B. Sanchez (Supervisor)
- Mike Filhiul (CPC 11)
- Randy Patchett
- Andrew LeBovioge
- Beth Richardson
- Jackie Christian
- Laurie Lolio (CPC)
- Ronnie McKinley (OS)
- Doug Mathews (CPC)
- Ruth D. Turull (OS)
- Eric Labardini (RSG NATCA Rep)
- Frank Whiten (OS)
- Philip S. Czervinske (ATCS)
- Fred Shufflen (SATCS)
- Ricky Higgins (SATCS)
- Frank A. Warford (OS)
- Kenneth J. Spoleti (ATCS)
- Jim Burnett (Supervisor)
- Steve Muenster (ATCS)
- Jack Overfield (STMC)
- David Tichavsky (OS)
- Charles Montgomery
- Carl Reed (OS)
- Walter Thompson (ATCS)
- Barbara Drewry
- John L. Price (SATCS)
- Bob Noeller

Continued on next page

TSD Familiarization Evaluation Report - ZHU, Continued

Special Factors

The training was conducted at three workstations on the control room floor. The ZHU site coordinator, Jerry Strickland, did an outstanding job in developing and implementing a timely schedule. Student experience with TSD varied from having worked with either TSD or ASD in TMUs, to having no experience. All students displayed a positive approach to the training. There were no coaching sessions on Wednesday due to the inaccessibility of TSD on the three nodes. The system was operational by 5:30 p.m. Wednesday afternoon. Training resumed Thursday morning as scheduled.

Evaluation Feedback

The following table lists the students' evaluation of the class, based on the responses of the 45 students.

Question	Average Response
1. My ability to use TSD has increased	4.64
2. Instructors clear and understandable	4.80
3. Instructors knowledgeable	4.82
4. Training will help in my job	4.62

Scores are on a scale of 1-5: where
1 = Strongly Disagree 2 = Disagree 3 = Neutral 4 = Agree 5 = Strongly Agree

Continued on next page

Appendix F: TSD Familiarization Evaluation Reports

TSD Familiarization Evaluation Report - ZHU, Continued

Most valuable part of course

The students responded:
- Interface with Rocky
- Learning new TSD functions
- Learning to use the TSD
- Being able to display the information
- Learning to display info
- Learning the basics about the TSD
- As a refresher
- Seeing the full capability of TSD
- Ability to pull up actual traffic density information (and chart)
- I'll know how to use if needed.
- Training others to use the display for information
- Hands on practice
- Learning what a TSD was
- Introduction to TSD functions and information - what is available
- Learning availability of drop down boxes - more user friendly
- Hands on after instructor brief
- Hands on training
- Ability to ask specific questions and observe actual inputs/instructions
- Having it
- All of it was good, need to know the basics to use machine well
- Learning about all of the options that are available
- Hands on time on TSD
- Hands on training
- Hands-on
- Learning how to access functions of TSD
- Hands-on part
- Hands on time
- Hands on
- Hands on with OJT
- Alert function
- Demonstration
- Basic knowledge and functions of inputs
- New information to apply to job
- Showed me where to find the information

Continued on next page

TSD Familiarization Evaluation Report - ZHU, Continued

What to include in future training?	The students responded: • Exceed software operation • More focus on the Exceed Program • N/A • More time to call up items • I'll let you know, but for now I think everything was covered. • Exceed software and how to display on overhead equipment • Won't know until I use the system more • Exceed Program - If this fails, we lose the ability to display. • Reports • Don't know at this time • The use of "virtual desktop" program • More time • ESIS Training • N/A • No comment
How could training be improved?	The students responded: • Exceed software operation • More focus on the Exceed Program • More time on practicing • Hold it away from actual operation; less distractions • Make it ongoing • Real time usage with Exceed software • More in-depth knowledge of TMU functions and database • Include more display tape info (i.e., the ability to display on the 4x8 blackboard) • Good job • It is good just the way it was presented. • More time (additional hour or two) • More time and hands-on • Not sure • Repeat it one week later. (But I understand our limitations.) • Not training at an operational position (designator) • Location, e.g., training department, however, might not be practical

Continued on next page

TSD Familiarization Evaluation Report - ZHU, Continued

Other comments

The students responded:
- Being able to see the "big picture" will enable us to understand some of TMU's decisions
- Great job! Thank you.
- Very well presented. As noted above, use of Exceed and visual display could be incorporated.
- Excellent instruction - just a lot to cover. Use of the system will bring more questions - one on one instruction was great.
- Everything went well. Other than the comments about Exceed.
- Very interesting!
- Slow down pace a little.
- Gail was wonderful. Very dynamic and helpful.
- Well prepared and presented.
- Excellent instructor
- Good job!
- Good tool.
- More sector specific demonstrations that would relate to a specialty.
- Questions # 6 and 7 could better be answered after I play with the equipment later.

TSD Familiarization Evaluation Report - ZNY

Dates	January 23-25, 2001
Location	ZNY- NY ARTCC
Population	30 students
Trainers	Gail Griffin, Dag Egede-Nissen

Students	Wolfgang Lerch (Ops Supv)	Michael Dawson (Ops Supv)
	Larry Jezouit (Ops Supv)	Tim Rosequist (Area Supv)
	Ken Dollard (Ops Supv)	Paul Stieglitz (Ops Supv)
	Steve Tsokris (OSIC)	Patricia Vega-Smith (Ops Supv)
	Robert D. Kavanaugh (OSIC)	Thomas Kelly (OSIC)
	Marvin Smolansky (Ops Supv)	Don Castonguay (OSIC)
	Fred Ashendorf (Ops Supv)	Paul Thumser (Ops Supv)
	Robert Giacomazzo (Ops Supv)	Peter Sheppard (Ops Supv)
	Ronald Bagley (Ops Supv)	Linda Corby-Leonard (Ops Supv)
	David Cohen (Ops Supv)	Michael Golden (STMIC)
	Jack Jackson (Ops Supv)	Kevin Delaney (Manager ZNY-505)
	Lawrence Bogner (Ops Supv)	Glen Cummings (ATCS NATCA)
	Carl Schroder (SATCS-Area D)	Barry Leach (Ops Supv)
	Steven Palmer	George Stewart
	Lois Esposito	George Leonard (Ops Supv)

Special Factors

The training was conducted the first day on a PC in a training office separate from the work bays. The PC was linked to a working node. Training was accomplished in an adequate manner; however, students with previous TSD experience remarked to instructors that the PC was slower than displays in the work area. Also during part of the day, a telcon was held in the same office while training occurred.

On the second day of training, the PC lost connection to its linked node. Site coordinator Paul Fairley moved training to an unused work bay and instruction continued with minimal interruption. The following morning, Bill Klare from the ZNY-504 office re-established the PC-node link and training was performed on the PC again.

Two students did not turn in evaluation forms. Lois Esposito asked for additional training on the following day, but did not appear for that training. George Leonard asked for additional time to complete the form in order to reflect on his responses and did not return to hand in the form.

Continued on Next Page

TSD Familiarization Evaluation Report - ZNY, Continued

Evaluation Feedback

The following table lists the students' evaluation of the class, based on the responses of 28 students.

Question	Average Response
1. My ability to use TSD has increased	4.68
2. Instructors clear and understandable	4.75
3. Instructors knowledgeable	4.75
4. Training will help in my job	4.46

Scores are on a scale of 1-5: where
1 = Strongly Disagree   2 = Disagree   3 = Neutral   4 = Agree   5 = Strongly Agree

Most valuable part of course

The students responded:
- Refreshing my memory of TSD functions
- One on one training
- Getting a good knowledge of TSD
- Hands on training
- The hands on portion of the training and the presence of an on-site instructor
- Sound fundamental instruction in TSD basics. I had no prior experience with program and was apprehensive
- Learning basic TSD commands
- Hands on live system
- Refresher type (I had spent 6 years in TM using ASD)
- Learning to use this equipment in a timely manner
- The in depth explanation on how to set my preferences
- Hands on training
- Hands on
- Learning how to use the different functions hands on
- The ability to "look into the future" for traffic volume and plan accordingly
- Hands on training
- Served as a refresher since I have been using TSD for past several years
- Hands on, and having knowledgeable instructor
- Hands on practice w/ instructor
- Basic operation of TSD
- Hands on – all the options

Continued on Next Page

TSD Familiarization Evaluation Report - ZNY, Continued

Most valuable part of course (continued)	The students responded: • Hands on • One on one explanation of TSD operation and capabilities • The whole thing
What to include in future training?	The students responded: • Flight projections for oceanic use (Domestic departure lists, etc) • Refresher • A little more time to work with the LIST request functions • Easy to read command reference manual • More specific training regarding LISTS • LIST reports detail • Too soon to say. Need to use it to see what else to ask. • How to change alerts dynamic on each machine • More about requesting LISTS and how to tailor those lists for specific info • Future upgrades • More specifics • No suggestions
How could training be improved?	The students responded: • Environment. Should be held in closed area – classroom. • ITS excellent • Each person have their own TSD while instructor instructs. • Additional time • Separate office to minimize potential distractions • More time available for training • Put individuals with TM (ASD) experience in class with similarly skilled individuals and people that never used the equipment with others who never used the equipment • Use equipment that doesn't have a ridiculously slow speed (like in ZNY-504's office). It hindered learning. • Should be moved to floor so that we could use a faster system. • Continue training updates later on. • More correlation between TSD/ETMS and the NAS – more site specific instruction

Continued on Next Page

TSD Familiarization Evaluation Report - ZNY, Continued

How could training be improved?
(continued)

The students responded:
- Free coffee
- No suggestions
- How about some breaks with a lengthened time period and more exercises.

Other comments

The students responded:
- Should be a 4 hour course
- No recommended changes at this time. Ms Griffin was very helpful and allowed us to thoroughly experiment with inputs to insure understanding
- Thanks
- Keep classes to one or two to keep learning going!
- The pace of training was slowed due to a slow system
- The training was excellent. Realizing there is a lot more to the capability of the equipment, it is important not to throw too much info at once. The pace was excellent.
- Training very well presented and easily understood. Gail presented information and course content in an excellent manner.
- The training was very good and well constructed.
- Add a little more time. Otherwise well done.
- Thanks. Good job.
- Great job!

TSD Familiarization Evaluation Report - ZID

Dates	February 6-8, 2001
Location	ZID - Indianapolis ARTCC
Population	52 students
Trainers	Joe Jankowski, Gail Griffin, Dag Egede-Nissen

Students

John Carter (CIC)	Bob Kuh (OS)
Russ Hansen (Area 4 – Supervisor)	Gary Cannady (ATCS)
Mike Looney ((SATCS)	Keith Krumwiede (ATCS)
John Kendall (CPC)	Brenda Woodard (ATCS)
Doug Kelley (O/S)	Brian Lucas (Ops Supervisor)
Andy Plonski (Area Supervisor)	Theresa Adams (ATCS/CIC)
Linda Povinelli (Ops Supervisor)	Christina Garcia (SATCS)
Bruce Mayo (Sup)	Tom Dury (Supervisor)
Bruce Aspley (Ops Supervisor)	Mark Batic (CIC)
Dwight Baker	Tim Kunkel (CIC)
Rich Beaman (CIC)	Jeff Bruns (ATCS/CIC)
Jay Wilkerson (CIC)	Marty Hanson (ASIC)
David Crittenden (Supervisor)	Greg Collins (Supervisor)
Charles McGrady (STMC)	Robert Plummer (Supervisor)
Ron Bishop (SATCS)	Rod Riggs (SATCS)
Richard Brown (Area 4 Supervisor)	Kevin Thomas (Ops Supervisor)
Alan Fogg (ATCS)	Gary Duemling
Larry McDowell	Tammy Burroughs (SATCS)
Ron Cunningham (SATCS)	John Lurker (SATCS)
Rodney Campbell (SATCS)	David Derrickson (CPC)
Kevin Wright (ATCS/CIC)	James Saunders (ATCS)
Jim Mulder (Ops Supervisor)	Doug Elkins
Larry Zan	Chris Decker (OSIC)
Darrell Hudson (SATCS)	Dave Kenney
Jeff Willingham (SATCA)	Thomas J. Allison (Ops Supervisor)
Jeff Cox (CIC)	Al Zielke (OS)

Continued on next page

TSD Familiarization Evaluation Report - ZID, Continued

Special Factors

The training was conducted on three PCs, located in a training classroom separate from the work bays. The PCs were linked to a networked workstation. One training session (4-6 AM on Feb 8) was conducted at the workstation site.

Training was somewhat affected by the TSD setup. The TSD display behaved as if the *Allow Primary Windows on Top* option had been set in **Style Manager**. That is, when the primary window was activated, all secondary windows slipped behind the primary and were hidden from view. Instructors were unable to change the Style Manager setting because its icon had been removed from the ETMS toolbar by the system administrator. This matter was discussed with Barbara Roddy, the System Administrator, who will verify settings and consult with TSD hotline if the difficulty persists.

Several supervisors had received on-site TSD training in 1999, while working as traffic managers in the TMU. One supervisor indicated that he had understood that the purpose of the present training session was to provide information in the use of Exceed to display a projected TSD image.

Evaluation Feedback

The following table lists the students' evaluation of the class, based on the responses of the 52 students.

Question	Average Response
1. My ability to use TSD has increased	4.67
2. Instructors clear and understandable	4.90
3. Instructors knowledgeable	4.87
4. Training will help in my job	4.65

Scores are on a scale of 1-5: where
1 = Strongly Disagree   2 = Disagree   3 = Neutral   4 = Agree   5 = Strongly Agree

Continued on next page

TSD Familiarization Evaluation Report - ZID, Continued

Most valuable part of course

The students responded:
- The whole thing
- Alert functions. Reading D/B's for reroutes.
- Hands on time. Small work group
- Flight list information. Hands on
- Learning the capabilities of the TSD
- How to request different Destinations to be displayed on TSD
- Hands on instruction
- Dag
- Hands on
- The use and benefits of TSD, and how it will help me do my job
- All questions answered clearly
- Hands on – practical applications
- The individual training. Having only 3 people in the class enabled us to ask detailed questions
- Hands on – practical exercises
- One on One/ Face to Face Training. The ability to ask questions
- Hands on instruction
- The chance to have hands-on experience vs lecture
- "Hands on" instruction
- Only having two of us trainees. It was much simpler.
- Hands on
- Hands on instruction
- Hands on time
- All aspects
- Review of some seldom used features
- The ability to have hands on and ask questions
- Hands on training
- Gives a good all round starting point to understand TSD capabilities
- Hands on time
- Hands on time is very beneficial
- Explaining the data
- Enables me to use a valuable tool
- Hands on use
- Interaction with a live instructor
- Hands on the computer

Continued on next page

TSD Familiarization Evaluation Report - ZID, Continued

• Most valuable part of course **(continued)**	• Hands on • For me, learning how "Exceed" was going to make the TSD work • All of it, because I had no knowledge about TSD to start with • All the training was valuable • Hands on training • Hands on • Hands on • Ability to configure TSD to multiple area needs. Individual instruction was good for me to learn. • The hands on time. I have been exposed to some TSD functions, but this gave me much needed background info. • Learning the various commands and how to personalize the TSD for myself • Knowledge of TSD as it can be applied to area of operations • One instructor/three students allowed for good hands on and appropriate interaction • One on one instruction • Review • Hands on practice • Question and answer
What to include in future training?	The students responded: • Everything seemed well covered • Keyboard shortcuts • Replays • Not only how to use the TSD, but methods to implement it • Hands on time • Follow-up more through training: how do you get special lists and filters? • Maybe a refresher to keep us informed of things we don't always use • Maybe more of the advanced functions • Knowledge of flight sets

Continued on next page

TSD Familiarization Evaluation Report - ZID, Continued

How could training be improved?

The students responded:
- Do not overlap sessions
- More hands on time
- Great! No improvement. The handout is a great tool.
- Better lighting
- Have a place where we could practice
- More time for practice
- 1 on 1 instruction. 3 to 1 OK, but 1 on 1 better
- It would be great to be able to practice right after this training
- This was very good for basic information
- Don't have everything scripted
- It's excellent the way it is
- Great course. Instructor very knowledgeable. Can't think of a way to improve it.
- Include the manual for each student.
- Longer period

Other comments

The students responded:
- Instructor knowledgeable and helpful
- It's nice to have an expert from the outside do some training correct!
- Good instructor. Pace was about right. Thanks.
- Good pace. Nice job.
- Human interface is much more effective than internet/disk based training
- The face to face is great. Better than CBI or video briefing.
- I like the face to face instruction
- In my situation as a CIC, my actual time using the product might be limited. Follow-up/Refresher training would be practical.
- Appreciate the live instructor format. It allows for questions and adjustments by the instructor based on student input
- Excellent
- Good training. Some limitations to the system but are problems with Exceed. (Noticed by experienced TSD user)
- Very good instruction
- Excellent training
- Good training. We need this technology.

Continued on next page

TSD Familiarization Evaluation Report - ZID, Continued

Other comments (continued)

The students responded:
- Hands on/live training gives the ability to ask questions and see them explained instantly. Very good.
- Give us the ability to change colors in the legend.
- Good training from the instructor. Knowledgeable. Answered all questions well.
- Very good training!
- Two hour training was right amount of time.
- Very helpful training – perfect length of course.

# TSD Familiarization Evaluation Report - ZJX

Dates	February 6-8, 2001
Location	Jacksonville ARTCC
Population	50 students
Trainers	Christine Risko and Charles Mohr

Students

Chubby Motin	Chris Hennig	Kelly Kendall
Steve Wilett	David Conley	John Voss
Sandy Martin	Phillip Bovinette	Darla Pettyjohn
Dave Scott	William Lineberry	Cliff Hare
Bill English	Dan Wilmer	Gathern Gillespie
Harrison Johnson	Perry Rose	John Sanders
Sharon Hyzer	Linda Powell	Howard Bedee
Patty Grieger	Kevin Pettyjohn	Mario Bosour
David Randa	Doug Pearson	Thomas Vihable
D. Dehne	Zeilnhiefer	Doug Henning
Irving Washington	Michael Cornman	Jim Heath
Buck Jetton	Paul Knight	Anthony Buie
Don Gallion	Robert Johnson	Gene Maynard
Mike Tallman	Alan Huntley	William Tant
James Grieger	Herman Hudson	Charlie Franz
Ed Goodwin	Charlie Riviere	Charlie Phillips
Sally Bell	R. Zuper	

Special Factors

Trainers were available for 6 blocks of coaching each day—10 a.m., 1, and 4 p.m. Each TSD training session was preceded by 30 minutes of training provided by site personnel, consisting of: ESIS, Status Information Area (SIA), Weather & Radar Processor (WARP), and access information for CCmail 8.2. TSD Training was held in 2 areas on the control room floor: 1 in an expansion bay with the ESIS projection unit and 1 in the special projects area of the TMU. There were no problems to report for this site.

Dan Wilmer did an excellent job coordinating all aspects of the TSD training. Dan had a schedule of 3 students per session and made sure all students were available and on time for each class.

Continued on next page

TSD Familiarization Evaluation Report - ZJX, Continued

Evaluation Feedback

The following table lists the students' evaluation of the class, based on the responses of the 50 students.

Question	Average Response
1. My ability to use TSD has increased	4.48
2. Instructors clear and understandable	4.67
3. Instructors knowledgeable	4.74
4. Training will help in my job	4.39

Scores are on a scale of 1-5:   where
1 = Strongly Disagree   2 = Disagree   3 = Neutral   4 = Agree   5 = Strongly Agree

Most valuable part of course

The students responded:
- To become familiar with the TSD and its functions and the features available.
- Hands on
- "Flights" commands, tailoring display to meet immediate needs.
- Walking through each of the commands.
- The use of monitor alert.
- Hands on with TSD.
- Seeing the changes made since I was in TMU.
- Refresher
- OJT
- Hands on portion
- All good
- Hands on
- Hands on use of the system.
- I will be training those that don't get this course, so the most valuable was to see Chris methodically follow a checklist & her excellent teaching techniques.
- All fairly valuable.
- Hands on training.
- Ability to visualize actual operation.
- Learning all the functions.
- Hands on
- Hands on training! Only three trainees!
- Hands on training.
- Highlighting of flights

Continued on next page

TSD Familiarization Evaluation Report - ZJX, Continued

Most valuable part of course (Continued)	• Relationships to windows. • Overall understanding. • Being able to get hands on time with computer. • Learning to utilize the TSD in everyday operations, not just TMU scenarios. • Hands on training. • Hands on • Knowledge of TSD acquired. • Learning the ability to display actual traffic. • Explanation of system. • Awareness of what TSD can do. • Seeing all the options available. • Learning the menus for TSD. • Hands on.
What to include in future training?	The students responded: • Anything that was not covered • Needs a site specific on scripts. • Information concerning enhancements. • In depth discussions on how to specifically tailor displays. • More hands on time to explore. • Hard to say until I use it. • This is mostly a traffic management tool and could hinder air traffic if the controllers see too much. • Learning to use the ETMS. • Shows the different functions of the TSD. • Learning "Flight" procedures, i.e. display/find, etc.

Continued on next page

TSD Familiarization Evaluation Report - ZJX, Continued

How could training be improved?

The students responded:
- None - Training met goals
- Shorten somewhat
- Individual computer terminals
- Can't be, she did a great job!
- One short break included.
- No idea
- More time of hands on
- This is good as it gets.
- Longer duration of hands on.
- Provide some additional time in TMU to see how it is used in the TMU unit.
- Individual hands on time - but understand why this may not be possible.
- Expand time to have course - make it a half-day instead of a half hour course.
- See above (6) - Hard to say until I use it.
- More time to learn.

Other comments

The students responded:
- Re-enforce…Scripts will be available.
- Had training in TSD last year. This is the reason for answer for #1 (3 neutral)
- Training good. Very beneficial.
- Less time on reroutes.
- The slower students should be move to back of class.
- Pace was good. Covered all topics. Thanks again, you've been a great help to our facility. Dan
- Pretty good all round!
- Spend more time on a/c heavy sectors.
- Seemed to be informative, now we just need to practice.
- Good overview

Appendix F: TSD Familiarization Evaluation Reports

TSD Familiarization Evaluation Report - ZLA

Dates	February 12-15, 2001
Location	Los Angeles ARTCC
Population	38 students
Trainers	Christine Risko and Charles Mohr

Students

Jerry Stone	Mark Denton	Larry Scherer
Jim Miller	Don True	Dave Williams
Aaron Boxer	Art Barton	David Schwartz
Garth Kolestman	M. O'Connell	Baker
Terry Griffin	Bill Poole	Ron Emery
Tom Wood	K. Raulston	Dan Alford
Melvin Perry	Bob Neher	John Mann
Mark Cottrell	James McDougald	Johnnie Garza
Curtis Cole	Kevin Stark	Robert Gagliano
Stephen Bieltel	Dwight Virtue	Gilbert Burnial
Stanley Faulk	R. Drewlo	Anthony Skinlick
Howard Long	Tim Savage	Jay Martin
Gary Tomak	Thomas Lynde	

Special Factors Trainers were available for 6 blocks of coaching each day—8 and 10 a.m., 1 and 3 p.m. Training was held in 2 separate areas of the control room floor.

Continued on next page

TSD Familiarization Evaluation Report - ZLA, Continued

Evaluation Feedback

The following table lists the students' evaluation of the class, based on the responses of the 50 students.

Question	Average Response
1. My ability to use TSD has increased	4.66
2. Instructors clear and understandable	4.76
3. Instructors knowledgeable	4.82
4. Training will help in my job	4.68

Scores are on a scale of 1-5: where
1 = Strongly Disagree 2 = Disagree 3 = Neutral 4 = Agree 5 = Strongly Agree

Most valuable part of course

The students responded:
- To learn all the capabilities.
- Monitor alert, map displays
- Display flights and alerts
- Explanation of options re: equipment
- Time for discussion
- Display of filtered flights
- I was familiar with basics - learned how to fine tune needed information
- Overall traffic display ability for selected airports
- Using flight info
- Learning of the system
- Ability to display individual aircraft info quickly and easily
- Review of TSD capabilities and use of drop down menus.
- Hands on training
- Ability to find/display individual flights, display wx intensities, and do replays.
- TSD toolbar
- Going through the various scenarios
- Hands on opportunity
- Info of selecting flight into airports
- Good review from TMU days
- The whole lesson was a good review
- All stages were valuable

Continued on next page

TSD Familiarization Evaluation Report - ZLA, Continued

Most valuable part of course (Continued)

The students responded:
- The ability to get full explanations for any question
- Overview
- Revisiting the TSD functions. It has been years since I worked with the "ASD".
- Basic exposed to TSD
- Increased understanding of TSD functions and capabilities
- Hands on ability
- Overall understanding of the functions and being able to get hands on experience.
- Good basic intro
- This was my first training with TSD. All was valuable.
- Review
- Flights and Alert features
- Asking questions/hands on

What to include in future training?

The students responded:
- Ask after we have some experience
- ? Hands on will take care of most of it
- None
- Don't know
- For a 2 hour block this training was complete
- Good content
- More about how to make your own script
- Will have to use the system before that can be determined
- None
- The info presented in the 2-hour session was adequate for a quick overview.
- Additional exercises on all aspects of TSD
- Spend additional time with examples and real time applications
- Any updates to software
- Creating scripts
- How to use in more detail the info that predicts saturation. What specifically would we do at ZLA with TMU to strategize solutions?

Continued on next page

TSD Familiarization Evaluation Report - ZLA, Continued

How could training be improved?

The students responded:
- No suggestion
- Can't see any improvement
- Don't know
- No suggestions
- See #6 (for a 2 hour block this training was complete)
- Have manuals for each
- User manual for the area
- For those not familiar, maybe 2-4 hr training with non-graded test to evaluate.
- No comment
- Same as 6 (will have to use the system before that can be determined)
- No comment as this time
- Offer coffee and donuts.
- Offer to all CIC's
- It was excellent as is
- One on one
- Self guided tutorial
- Extended amount of time allotted for each training session.

Other comments

The students responded:
- Very helpful
- Perfect as is
- 2 hrs. sufficient
- Pace was fine for the 3 of us with TMU time
- Generally seemed okay
- Good overview; use will increase my ability to use equipment
- Good session!
- This was one of the best training sessions due to the fact there was pertinent information without delays and time to apply.
- The data was presented in a very understanding manner and all questions were addressed.
- Thanks for the help. Good job.
- Develop training problems
- Good job using real equipment.

TSD Familiarization Evaluation Report - ZTL

Dates	February 20-22, 2001
Location	ZTL - Atlanta ARTCC
Population	45 students
Trainers	Justyne Johnson, Charles Mohr, Dag Egede-Nissen

Students

- Frank McDougal (ATCS)
- Richard McCarthy
- Michael Muhammad (OSIC)
- Jerry Rutherford (OSIC)
- Richard Moser
- Michael King (OS)
- Don Harkins
- Jeffrey Graves (OS)
- Ken Dancy
- Gene Smith
- Tim Helms (SATCS)
- D. McMillan (OS)
- Jon Uphouse
- Greg Jones (OS)
- Ron Wilder (OS)
- William Moore (OS)
- James Thomason
- Laura Guarracino
- Norman Neeley (OS)
- Malanie Jones
- Tom Hayes (OS)
- Tom Titshaw (CIC)
- Robert Moran (Ops Mgr)
- Robbie McGrath
- Mark Schoenhoff (ATCS)
- Mike McKine (Sup)
- Jose I. Rivera
- Richard M. Martinez (SATC)
- Randall L. Baughcum (ATCS)
- Michael Commander (OM)
- Debby Dennis (OS)
- Harry Hyaduck
- Bob Woeste (CIC)
- Jerry H. Hodge (OS)
- Fred Gleason
- Ronnie Nobles
- Michael Harlow
- Marv Horscher
- Robert J. Walters, Jr. (OS)
- Paul Holes (ATCS-CIC)
- Jane Gallo (OS)
- Scott H. Siegel (OSIC)
- Doris Wilson (Ops Mgr)
- Darryl k. Dudley
- Joel White (OS)

Continued on next page

TSD Familiarization Evaluation Report - ZTL, Continued

Special Factors

The training was conducted on three PCs, located in a work bay reserved for training. The PCs were linked to a networked workstation. Three training sessions were conducted at the same time daily: 9:15-11:15, 12:45-14:45, and 15:00-17:00. The three PC's were situated next to one another in close proximity, which caused a degree of congestion when nine students received training. When six or fewer students were present, the training facility did not seem crowded.

Evaluation Feedback

The following table lists the students' evaluation of the class, based on the responses of the 45 students.

Question	Average Response
1. My ability to use TSD has increased	4.58
2. Instructors clear and understandable	4.80
3. Instructors knowledgeable	4.87
4. Training will help in my job	4.69

Scores are on a scale of 1-5: where
1 = Strongly Disagree 2 = Disagree 3 = Neutral 4 = Agree 5 = Strongly Agree

Continued on next page

TSD Familiarization Evaluation Report - ZTL, Continued

Most valuable part of course

The students responded:
- Hands on
- The command functions
- Everything. Use of maps, flight displays, wx, showing lists of alerted sectors.
- The flexibility to customize the TSD to the particular needs of the area/individual
- Learning how to pull LISTs
- Understanding the commands and how to manipulate the TSD display
- Shortcuts to display
- Learning how to use each item on the display
- Instruction
- "Hands on". Actually working with live equipment
- Actually having hands-on time with the equipment
- I use it every day. A great refresher
- Good refresher
- The hands on time was most valuable.
- Hands on training
- Experiencing different options on the TSD
- Hands on training
- All
- All information was new
- Going over each function separately
- Every bit of it
- Good review of things I've found through trial and error
- Display functions
- Learning to request reports
- Hands on time
- Learning to bring up sector activity
- Hands on; drop down menu command(s) explanation
- Hands on
- Show flights section was most valuable
- Hands on
- Being able to ask questions
- Learning how to retrieve and display information
- Being able to try things in a hands on setting

Continued on next page

TSD Familiarization Evaluation Report - ZTL, Continued

Most valuable part of course (continued)
- Hands on training
- Creating personal scripts
- Learning how to use a tool that will make me more effective
- Learning the many diverse uses of the TSD
- Learning what you can do and how to do it
- Learning how to display flights and weather
- All
- Hands on instruction
- Hands on time
- Hands on training

What to include in future training?

The students responded:
- Specific training on tools within the TSD
- A little more training on how to customize tools
- How to set up tabs/tool bar
- None
- Other features
- Covers all
- Nothing I can think of
- More in depth information concerning report parameters
- A repeat
- This was very comprehensive
- Can the size of the icons be changed?
- More information regarding the "Command Line"
- None at this time

Continued on next page

TSD Familiarization Evaluation Report - ZTL, Continued

How could training be improved?

The students responded:
- I see nothing in need of improvement
- Very good as is
- I cannot think of anything. It was very thorough.
- Leave more time with instructor. Cover more material.
- Maybe something particular to individual areas.
- It was good as it is.
- Good overview of system. No suggestions.
- No suggestions.
- A workstation for each of us. One for three.
- Maybe go into a little more detail and increase time to train on it.
- Less time on weather.
- More hands on time to try different things.
- Make it 1 on 1 and conduct training in a better environment, where we are not listening to other's training.
- Don't have to.
- More time.
- More time.
- Get it out of the control room and one terminal per person.
- No recommendations.

Continued on next page

TSD Familiarization Evaluation Report - ZTL, Continued

Other comments

The students responded:
- More time would be nice
- The amount of time was just right and the information covered was very adequate
- More time to do it ourselves!
- All OK
- Slow down or add more topics. A beautiful tool.
- Add airport runway lengths to display.
- Slow down. This could easily cover 8 hours.
- Good overview of TSD.
- Pace was excellent.
- Dag is very knowledgeable and keeps your interest.
- It was good. Might could have moved a little faster.
- I believe the information was completely adequate to allow me to learn on my own any other particulars.
- It is hard to say until I've had a chance to sit and play with it.
- Less time on using asterisk and question marks to find flights.
- More time.

TSD Familiarization Evaluation Report - ZAU

Dates	February 27 - March 1, 2001
Location	Chicago ARTCC
Population	66 students
Trainers	Christine Risko, Gail Griffin and Charles Mohr

Students

Michael Kline	Ellison Hernann	Robert Szajkovics
Steve Roseman	Gary Knapp	Larry Eden
Henry Farmer	Wayne Smuda	Robert Sliwa
Michael Borzym	Kevin Strand	Dave Voggesser
Carl Knapp	Gerald Heard	Scott Bronger
Eric Widick	South Area Sup.	Keith Stokes
Tom Reisel	John Dittrich	Terry Starck
John Belts	Ralph Nerino	Mark Dombrowski
Bill VanLoan	Wendy Cook	Diana Mulka
Thomas Rucker	Kent Karr	Greg Weeks
Wayne Winslow	Dave Merkel	Dennis Shanks
Gelaine Gallucci	Tim Plezbert	Jackqueline Whitaker
Milan Indrisek	Patti Policht	Peter Sober
Jennifer Ceithaml	K.J. Friedlein	Steve Ford
Sally Sprengel	Rich Byrne	Socrates Passialis
William Wichial	Dennis Jarnecke	Bill Meyers
Gregory Wyse	Terry Gilbert	Gordon Broich
John Gurley	Bill Cook	Bill Grimm
Barb Schennum	David Scaffidi	Rod Carver
Sean Hathaway	Andrew McMullen	Pat Siwak
Dennis Ithal	George Tsuchiyama	Jeff Spang
Chris Bryja	Davied Ingraham	David Prochaska

Continued on next page

Appendix F: TSD Familiarization Evaluation Reports

TSD Familiarization Evaluation Report - ZAU, Continued

Special Factors

Trainers were available for 9 blocks of coaching each day—2 trainers at 8 a.m., 10 a.m. and 1 p.m. and 1 trainer at 2, 4 and 7 p.m. Students had expected us to provide ESIS projector training and were surprised that we were providing only TSD training. Training was held in the computer room in the basement of the ARTCC. The training environment which we were provided caused several students to question its suitability as a learning environment:

- The temperature was extremely cold; there was a sign in the control room telling students to bring their coats with them. (Just about every student complained about the temperature).
- The equipment was noisy which made communicating with the students difficult,
- There were 3 computers set up in the middle of an open space between racks of equipment; when we had 3 classes with 3 instructors and 12 students, we were bumping into each other,
- There was a notice on the door before entering the room stating there was "friable asbestos" present

Evaluation Feedback

The following table lists the students' evaluation of the class, based on the responses of the 66 students.

Question	Average Response
1. My ability to use TSD has increased	4.48
2. Instructors clear and understandable	4.67
3. Instructors knowledgeable	4.74
4. Training will help in my job	4.39

Scores are on a scale of 1-5: where
1 = Strongly Disagree 2 = Disagree 3 = Neutral 4 = Agree 5 = Strongly Agree

Most valuable part of course

The students responded:
- Learning to use the equipment.
- Reports and alerts = help prepare for heavy periods. Combine/decombine.
- I learned how to display flights that affect my area.
- Introduction and explanation of equipment.
- Maps and depictions of my area of specialization.
- Hands on
- The hands on time.

Continued on next page

TSD Familiarization Evaluation Report - ZAU, Continued

Most valuable part of course Cont'd

- Refamiliarization with TSD.
- Select flights and alerts.
- Learning how to use TSD.
- To refamiliarize myself with TSD (previous TMC) for flight display characteristics
- Better familiarization with system.
- All of it - learning how I can use it in day to day work
- Entire training
- Learned the basic use of TSD for CIC duties.
- Hands on
- The operation of TSD software
- Direct application to traffic management - flight filtering
- Being able to work the machine
- Just the ability to use the equipment
- Setting up the TSD, for when I have CIC duties.
- Hands on
- Very detailed
- Review of TSD functions
- Active participation
- All parts will apply - WX displays/alerts/flight info all will enhance the operation.
- "Hands on" demonstrations.
- Hands on practice
- Playing with the TSD
- Familiarization
- Reports msgs
- The hands on - creating displays
- The weather display should help control, the other stuff I'm not sure
- Review of functions for keyboard
- Receiving it
- All
- The hands on experience. Much easier to comprehend equipment
- Understanding overall TSD uses
- The hands on experience.
- Seeing some of the functions that will be useful on the floor.
- It will enable me to use the equipment already in use.
- Hands on
- Information
- Actually, I probably will not use this often, but it's nice to know.

Continued on next page

TSD Familiarization Evaluation Report - ZAU, Continued

Most valuable part of course Cont'd

- Just being able to get hands-on with the TSD.
- Review of TSD functionality
- Knowing exactly how it's supposed to work rather "Well, this works, but I'm sure there's a better way."
- Features that were available and how to extract the data.
- All of it
- ESIS usage
- Will benefit my ability as a CIC
- Learning how to develop and save displays
- Enhanced knowledge of tools available to a SUP/CIC for flowing traffic.
- The information was presented in a very logical sequence, making it easy to follow.
- Being able to display when sectors will be busy.

Continued on next page

TSD Familiarization Evaluation Report - ZAU, Continued

What to include in future training?

The students responded:
- ESIS training
- Impact on moving aircraft from monitor alerts sectors to "alleged" slower activity sectors.
- There is enough now to start with
- This was fine.
- How to use ESIS.
- Bring in some information on the ESIS projector system
- Refresher
- Not sure
- None
- Most seemed to be covered quite well
- Not sure, maybe better fonts.
- None - very comprehensive
- ESIS operation
- None right now - need more familiarity with the tool to determine.
- None
- Can't think of any right now
- Short cut keys outlined
- ESIS
- None
- Shortcut command card
- ESIS information and usage
- ESIS display
- More practice (maybe have 5 things or scenarios you need to complete)
- None
- ESIS usage
- Don't know yet
- Sector saturation demo and relief of SAIs

Continued on next page

TSD Familiarization Evaluation Report - ZAU, Continued

How could training be improved?

The students responded:
- More comfortable environment, the location was extremely cold.
- Warmer environment.
- Classroom display with overhead depictions.
- Help my memory
- Have a warmer room.
- Excellent as is
- Not sure
- Held in a warmer room
- Fine job in training
- Warm up the room
- Training tied into ESIS applications.
- Longer time period for same info - more practice time.
- Better environment
- Maybe a little more time (2.5 hrs)
- More hands on for individuals
- Don't know
- A warmer area to hold the training in
- Practice
- Foot heaters
- Warmer location
- Quieter location
- I wanted training on the ESIS equipment, not solely the TSD
- Shorten it to 1 hour for the tools I'm likely to use
- Quieter room
- Good working basic knowledge - kept it simple
- In a warmer room
- Tell people what they are going to learn ahead of time
- Up in a classroom, not the Dungeon!
- Not sure
- It couldn't
- How do you use ESIS?
- Allow more practice. Have an independent system that is not affected by center traffic load.
- More hands-on
- Train on the floor

Continued on Next Page

TSD Familiarization Evaluation Report - ZAU, Continued

Other comments

The students responded:
- Very well done!
- Have 2 individuals from an area train at the same time to start creating an area database that area personnel are interested in.
- Training on overhead display is sorely needed, i.e., overhead projectors.
- The presentation and instruction was given at a good pace.
- Excellent pace of instruction!
- Poor training environment (in basement) Very COLD!
- Good job - worth the time - helped greatly - better that the TSD CBI
- For a non-TMU background supervisor, I thought it was excellent Thank You!
- More hands on time
- Make sure the machines have enough resources to handle the software.
- Training was just the right amount of time.
- Just about right
- The training was very good. Info was complete and very usable.
- Good pace, good instructor
- More hands on
- Excellent job presenting TSD operating capabilities.
- Longer time period for same info - more practice time.
- More time on weather
- Very good training for beginning TSD. Thank You
- Very good class
- Just the right amount of time to cover what is needed
- Most things shown will never be used by me or most supervisors.
- Very Good. Training room could have been better
- I was told I was going for ESIS training and ended up it was TSD training.
- Add more ESIS training
- Perfect pace, good course
- Great job by instructor, very knowledgeable!
- Add ability to call up airport and have available on one screen - runway/lighting/navaids/weather - in emergency situations
- Trainer was excellent.

TSD Familiarization Evaluation Report - ZAB

Dates	March 6-8, 2001
Location	Albuquerque ARTCC
Population	26 students
Trainers	Christine Risko and Tom Alizio

Students

Ben Daniel	Dennis Roybal	Brenda Perry
Rick Chavez	Dave Mott	Chris Abeyta
Lee Silva	Linda Brown	Trey Madrid
Mark Spaulding	RA Sutton	T. Baecher
Jeff McElwain	Larry Rolls	Greg Tingley
Gary Postlewait	Richard Acker	Richard Boatman
Michael Szucs	Don Scholosser	Carol Latham
Vincent Abeyta	Duane Hillerson	Janet Mould
Andrew Rankin	Melinda Drumheller	

Special Factors Trainers were available for 6 blocks of coaching each day—11:30 am, 2:30, and 5:30 p.m. Training was held in 2 separate areas of the TMU.

Evaluation Feedback The following table lists the students' evaluation of the class, based on the responses of the 26 students.

Question	Average Response
1. My ability to use TSD has increased	4.42
2. Instructors clear and understandable	4.65
3. Instructors knowledgeable	4.73
4. Training will help in my job	4.38

Scores are on a scale of 1-5:   where
1 = Strongly Disagree   2 = Disagree   3 = Neutral   4 = Agree   5 = Strongly Agree

Continued on next page

TSD Familiarization Evaluation Report - ZAB, Continued

Most valuable part of course

The students responded:
- Sizing and moving maps - information that can be derived from reports.
- Hands on
- All
- Setting up my own file
- Hands on training
- Use of the alert function
- Learning more detail in how to use the TSD. Examining sectors.
- The instruction on how to use the TSD vs. prior "on-the-fly instruction.
- Hands on training on use of TSD.
- The familiarity.
- I use this everyday.
- Learning different functions not known before.
- Being able to save my own presentation.
- Set up for area and customizing information.
- Starting the TSD
- Utilizing information available.
- I was not familiar at all with ETMS, so the whole class was valuable.
- Learning how to set up the display.
- Examining lists.
- Hands on use of TSD.
- All of it.
- Hands on
- Learning new options.
- Hands on experience - learn by doing it.
- Ability to display more than 1 TSD with use of ESIS at a time and ability to develop a reroute.

Continued on next page

TSD Familiarization Evaluation Report - ZAB, Continued

What to include in future training?

The students responded:
- None
- Good job
- None
- None that I can thing of.
- None
- ESIS training for use with overhead projectors.
- None
- ESIS projector and processor information.
- More WX
- How to put screen displays on wall displays.
- How to save alerts to your file and have those alerts come up when you open your file. (NOTE: Multiple alerts timelines)
- None
- Add general preventative maintenance practices for AF Techs (NOTE: AF Tech who sat in on class to become familiar)
- Unsure at this time

How could training be improved?

The students responded:
- Follow-up after we've had an opportunity to practice what we've learned.
- Good job
- More quiet place, less distractions
- Specialty specific setup.
- Unknown
- Concurrent with installation of equipment.
- No idea
- More examples.
- Couldn't - it was excellent!
- Follow up with refresher training.
- Add general preventative maintenance practices for AF Techs (NOTE: AF Tech who sat in on class to become familiar)
- Hold it in a tavern.
- No change

Continued on Next Page

TSD Familiarization Evaluation Report - ZAB, Continued

Other comments

The students responded:
- This training was very good!!
- This training should have been given prior to installation of TSD at AOS positions.
- Thanks
- Good content and adequate time allotted for training.
- 2 students in training (NOTE: class had 3 students)
- The pace of the class was good.
- Very professional and detailed - Thanks.
- Well done.
- Great presentation! Thank you.
- No change

TSD Familiarization Evaluation Report - ZDV

Dates	March 13-15, 2001
Location	ZDV - Denver ARTCC
Population	43 students
Trainers	Charles Mohr and Gail Griffin

Students

- Robert Roane (OS)
- Tony Knudson (AF Tech ESIS)
- Thembi Ndlovu-Hickey
- Norman K. Tate (OS)
- William E. Jones
- Rick Shively
- Scott Steinbrecher (Supervisor)
- Jim Rinehart (CPC)
- Jack Shisler (OS)
- Randall Schlecht (OS)
- Gregory Welz (ATCSS)
- Joseph C. Coenen (CPC)
- David W. Norris (CPC)
- Shelly Dambeck
- Ted Lucero (Maintenance)
- Brian Hall
- David Zimmerman
- Peter Hommertzheim
- Steve Ruttman
- Alex Apperhaus
- Steven Deubler (OS)
- Wayne Patrick (OS)
- Patrick Newton (Controller)
- Paul M. Bonestroo (ATCS)
- David Richardson (System Specialist)
- Gordon Bertoglio (Automation Tech)
- Donald E. Hickman Jr. (ATSS)
- Melody Hanner (CPC)
- Kelly Neuhaus (CPC)
- Mark Robb (CPC)
- Tim Deckert
- Gary Butler
- Barry Garlick
- Ron Galbraith (CPC)
- Kevin Wright (OS)
- Dan Fedorowicz
- Roger Nakata
- Jean E. Needham (OS)
- Scott Nuffer (ATCS CIC)
- John Ascher (ATCS)
- Richard Freigain (OS)
- Ronald Goecke (OS)
- Glenn Contouris (CPC)

Special Factors Two PCs located in the DYSIM were used for training. Nine students from automation/maintenance attended the training sessions.

Continued on next page

TSD Familiarization Evaluation Report - ZDV, Continued

Evaluation Feedback

The following table lists the students' evaluation of the class, based on the responses of the 43 students.

Question	Average Response
1. My ability to use TSD has increased	4.47
2. Instructors clear and understandable	4.70
3. Instructors knowledgeable	4.65
4. Training will help in my job	4.35

Scores are on a scale of 1-5: where
1 = Strongly Disagree 2 = Disagree 3 = Neutral 4 = Agree 5 = Strongly Agree

Most valuable part of course

The students responded:
- Moving data blocks and using filter command
- Understanding AT use of ESIS
- The training provided a good overview. It taught some good shortcuts and tidbits.
- Examining alerts and obtaining info to make decisions
- Being able to get around the tasks
- Hands on
- Hands on training
- Functions I was previously unaware of are now understandable
- Navigating through the system with hands on training - great
- Any training is valuable
- Learning the different capabilities of ETMS - and the commands
- Hands on with the TSD
- The overview of traffic flow and what ATC looks at
- Review of all data most useful
- Learning to write a script
- Hands on exercises
- Refamiliarizing myself with the capabilities of the TSD
- Going through all the menu bar items
- Showing weather and different designations of aircraft: H, J, T, P
- The step by step progression through each menu
- Hands on training
- Knowing how to use equipment
- Hands on experience

Continued on next page

TSD Familiarization Evaluation Report - ZDV, Continued

Most valuable part of course (continued)
- All of the training was excellent!
- Ask as many questions that I could think of
- Learned about numerous capabilities I didn't know about
- Actually get to use the equipment, not just classroom
- How to set up your own script and files
- Total overview with specifics
- Quick ID of aircraft - quick reference settings
- Learning how to create displays
- All of our time together
- Just learning the commands and the abilities of TSD
- Hands on
- The hands on instruction and teaching techniques of instructor
- Getting an overview of TSD. Getting questions answered

What to include in future training?

The students responded:
- Technical description
- I got all I needed for what I use TSD for.
- I am a tech. (Maybe some troubles)
- Add Exceed, if not for info purposes only have some minor helps.
- Can't think of any.
- Recovery if system locked up and no mouse active.
- Handout that would walk a person through the exercise, as being out of it for a week I'd forget most of it.
- None
- More time to talk with the instructor
- None
- Script writing
- TSD and ESIS interaction
- None
- Exceed and overhead projectors

Continued on next page

TSD Familiarization Evaluation Report - ZDV, Continued

How could training be improved?

The students responded:
- A break half way through
- Possibly more hands on. After the lesson use a handout with steps to perform functions.
- More of it, longer times, service aids
- Repetition or refresher
- As a technician, would like to see some troubleshooting problems
- Cover R Manager to look up fixes and airports - but not part of TSD??
- More terminals, individual training lessons to create items and play more
- Provide training prior to installation of TSDs in areas.
- Add some basic UNIX commands with regular menus.
- Handout that would walk a person through the exercise, as being out of it for a week I'd forget most of it.
- Maybe, more time to use equipment
- Allow more time.
- Don't know
- Get it prior to getting the TSD on the floor.
- Two sessions to reinforce what we practiced
- N/A
- No ideas on that - excellent!

Other comments

The students responded:
- It was interesting and enjoyable.
- More of it, longer times, service aids. Thanks for cheat sheet.
- Time allotted was good.
- Good for allotted time!
- Thanks
- Great class, much needed
- Pace OK for our group of 3
- No changes
- Good pace
- Pleasant instructor kept the class moving at adequate learning pace.
- Good class. Enjoyed the time spent.
- Charles presented this lesson very well.

Continued on next page

TSD Familiarization Evaluation Report - ZDV, Continued

Other comments (continued)
- Good job
- Small groups were very effective.
- Very close to a perfect amount of time and instruction. - Very helpful
- Would like hourglass symbol (similar to "Windows") when TSD is contemplating entry.
- Great, concise course.

TSD Familiarization Evaluation Report - ZFW

Dates	March 20-22, 2001
Location	ZFW Dallas/Ft Worth ARTCC
Population	67 students
Trainers	Gail Griffin, Charles Mohr, Dag Egede-Nissen

Students

Gary Andrae (CPC)
Ron Vick (OS)
Michael Olszewski (OS)
Mark Gordon (OS)
Thomas J. Powers (OS)
Dave Henderson (ATC)
Carl Youngblood
William Arnett
Mike Demboski (NATCA-Area rep)
Darryl Walker (OS)
Craig Lindelow (Training Spec)
Lynne Leverenz (STMC)
David Benedetto (CPC)
Gene Hutchins (OS)
Mark Kennedy (OS)
Bryan Henderson (OS)
Bruce Kane (SATCS)
Margie Juenemann
Russ Emmert
Ernie Valdez (OS)
Robin Crook (OSIC)
David Asbell (OS)
Eric D. Malmberg (OSIC)
Scott Weining (ATCS)
Rowland Ballard
Lori Berry (OS)
Dutch Daugherty
Richard W. Herbst (SATCS)
Vern Leder (OS)
David Ritchey
Frank Ward (CPC)

Jim Thornton (SATCS)
Michael T. McCully
Melv Netzer (SATCS)
Philip Booker (OS)
Matt McCorey
Buster Porter (ATCS)
DeAnn Martin (SATCS)
Helen Mahon
Deborah Walker (OS)

Thomas A. Sanderlin (SATCS)
Mark McConnell
Alan White (SATCS)
Wayne Coley (ATCS)
David R. Crowell
Mary Hokit (OSIC)
Jim Hanlon (SATCS)
John C. Tittle (OS)
Terry Lee
Nick Blaylock
Michael Roy (ATCS)
Timothy A. Smith (CPC)
Barry Gossett (OS)
Ron Laster
Edward Vaca
Ray Bosemer
Eric Arnold (ATCS)
JoAnn Hansen (ATCS)
Phil Layman (ATC)
Kevin Davis (Staff Specialist)
Geoff Pertzborn (OS)
Jamie Mannon (CPC)

Continued on next page

TSD Familiarization Evaluation Report - ZFW, Continued

Students (continued)

Ed Hulsey (CPC)
Ron Lozano (OS)
Steve Madick (OSIC)
David Powers (OS)
Chris Rice (ATCS)

Special Factors

The training was conducted on three PCs, located in a training room separated from the work bay area. PCs were linked to different networked workstations. Instructors provided two training stations at 6:30-8:30, 8:45-10:45, and 11:45-1:45 and one at 2-4, 4:15-6:15, and 7:15-9:15. The three PC's were situated next to one another in close proximity, but because trainers used no more than two stations at any one time, there was ample space for students.

A large number of ZFW personnel attended training. Nearly every session had either two or three trainees in attendance. At the 11:45 session on March 22, one instructor (Charles Mohr) accommodated six students.

The ZFW POC (Kevin Davis) had planned to conduct ESIS workstation training immediately prior to each scheduled TDS training activity, but due to technical difficulties, ESIS training could not be accomplished. He did leave the monitor and projector in place and requested TSD trainers to describe the system if asked. Generally, students who did ask about the ESIS set-up were satisfied with verbal descriptions, although a small number indicated that they wished that they could work with both windows (a TSD display on the monitor and another TSD display on the wall) during training.

TSD trainers commented on the exceptional work of POC Kevin Davis, who coordinated the scheduling of an unusually large number of attendees, all of whom arrived enthusiastic and at appropriate times.

Continued on next page

TSD Familiarization Evaluation Report - ZFW, Continued

Evaluation Feedback

The following table lists the students' evaluation of the class, based on the responses of the 67 students.

Question	Average Response
1. My ability to use TSD has increased	4.73
2. Instructors clear and understandable	4.84
3. Instructors knowledgeable	4.88
4. Training will help in my job	4.79

Scores are on a scale of 1-5: where
1 = Strongly Disagree 2 = Disagree 3 = Neutral 4 = Agree 5 = Strongly Agree

Most valuable part of course

The students responded:
- The hands on instruction
- Hands on
- Hands on instruction
- Hands on
- Hands on application/practice
- Dag was patient and had some knowledge of ATC to help him understand my requests
- Increase knowledge and use of TSD function and capability
- Hands on practice and ability to depict my particular area
- All parts about equal
- Hands on training with instructor
- Hands on time
- Gail. Excellent job. Thorough.
- Hands on practice
- Exposure to actual inputs
- Hands on exercises
- Good overview for CPCs
- Total review and specific application training for job
- Overview of ETMS software
- Hands on time
- Hands on
- Learning what can be displayed
- Hands on experience
- Hands on application

Continued on next page

TSD Familiarization Evaluation Report - ZFW, Continued

Most valuable part of course (continued)

The students responded:
- Hands on
- Hands on application
- Understanding the various commands
- Hands on
- Learning features of the TSD
- Learning how to select flights that will affect my area
- Hands on real time
- Hands on training. CBI was good, but it can't replace hands on time and one on one with instructor
- Hands on with equipment
- All of it
- Using actual information for training
- Learning how to select individual sectors and the traffic that will be entering them
- Hands on review of previous CBI
- Hands on instructor being right there
- Thorough instruction of all the menu functions will make it very easy to set up TSD in my area
- Q & A in real time
- It was all excellent
- The hands on training at the work station
- Hands on training
- Hands on part of lesson
- Becoming familiar with setting up the TSD to help manage traffic in my area
- The hands on. To see the different windows/options opened/used
- Hands on & instructor knowledge to assist with questions, concerns, and resolutions
- Yet to know
- Display of capabilities of the system
- Hands on time
- Hands on training
- Hands on application
- Learning how to pull up the traffic situations entering my airspace
- Hands on time
- To learn functions of ASD as a tool for OS
- Alerts

Continued on next page

TSD Familiarization Evaluation Report - ZFW, Continued

Most valuable part of course (continued)
- Hands-on training prior to use on the floor
- Multitude of info that can be displayed for flights and sector projections
- Hands on training with an instructor
- The one on one/hands on briefing. Gail does a very good job of presenting the material

What to include in future training?

The students responded:
- Setting up multiple windows (more) and site specific exercises
- More facility specific examples, ideas from ZFW controllers
- Any improvements made
- Adaptation of playbook into TSD training
- Ability to see alerts for geographic areas
- Everything was covered adequately
- Enhanced or additional features
- We did not use multiple windows, although I believe this would be simple.
- Too soon to know
- Just more hands on time
- Separate areas that TMU would normally need from areas that OS's will need
- Too early for me to tell. I still need time to experiment
- Try to be specialty specific
- Additional time for more practice
- Practice
- How to display on screens
- Use of scripts, commands
- Update training and refresher training on areas not used as much
- Everything was covered more than satisfactorily.

Continued on next page

TSD Familiarization Evaluation Report - ZFW, Continued

How could training be improved?

The students responded:
- The training was excellent
- Nothing – Good job!
- More play time
- Maybe and additional hour of self-exercise
- Develop displays tailored to individual area of specializations and demonstrate functions helpful to each area
- Until implemented on floor, no way to determine how to improve
- More hands on
- More hands on time
- Two to one ratio on TSD
- Not sure. It was quite adequate.
- Simulation. More instruction.
- Additional time for more practice
- Training to be conducted by specialty
- Worked out great
- Have a refresher CBI from time to time
- One terminal per student
- Not all functions were working properly
- It was great!
- Each person using an individual computer to follow instruction would be much better than rotating through a single computer.
- All the essentials were covered
- Extend length of training so all operational employees could attend
- More computers to allow for more hands on time
- A little more practice
- More hands on
- Study guide to go with class
- CBI unnecessary. Everything covered during hands-on training.
- Supply donuts

Continued on next page

TSD Familiarization Evaluation Report - ZFW, Continued

Other comments

The students responded:
- Good job!
- The training pace was perfect. Gail was very helpful in answering specific questions.
- Excellent instruction
- Slow down pace. Glad to get menu functions
- The training was very worth while
- Instructor was very thorough. Great job!
- Time limit is very good. Instructors are very prepared and able to answer all questions.
- Come back 6 weeks after system implemented to catch any/all problems
- Overall covered everything. Maybe a follow-up class 3 months after implementation
- Good job teaching the system
- The instructor was very knowledgeable and helpful
- The length of time allotted during the training session seemed adequate for covering the basics of TSD
- Very informative. This (TSD) will be a very useful tool to both the controllers and supervisors
- Thanks
- Everything seemed to be just perfect. If it was longer, it would get boring. If it was shorter, instructor would have to rush.
- The one on one aspect was very good instead of sitting through CBIs that are boring.
- Excellent job
- Instructor was very enthusiastic and motivated. Video card errors every now and then.
- Good instruction. Charles was knowledgeable and answered all my questions.
- Instructor was excellent. Good pace and amount of information
- Good job
- It was informative and beneficial. Consider longer individualized training.
- Gail – Excellent job – Good instr – More time for training could help when we use the data –We probably need to develop a higher kind of XXX (illegible writing) during actual operations.
- Maybe more work on lists, such as counts every 5 minutes, etc.
- This equipment does not appear to aid the controller to do his job.

Continued on next page

TSD Familiarization Evaluation Report - ZFW, Continued

Other comments (continued)

The students responded:
- I had no previous knowledge of TSD functions. Now I feel I have a good overview.
- I enjoyed the course. Thank you.
- This training gave a great initial overview of the ETMS functions
- Thanks
- Just right
- Nice program
- The instruction was thorough and very informative. Speed of instruction was good and instructor was very open to questions and did not hesitate to answer any and all questions.
- Very useful training, presented in an easy to understand fashion.
- One on one training was terrific. It will help me perform my BYP team functions much better concerning TSD and ESIS. All questions were answered fully.

TSD Familiarization Evaluation Report - ZMA

Dates	March 20-22, 2001
Location	Miami ARTCC
Population	43 students
Trainers	Justyne Johnson and Joe Jankowski

Students

Michael Byrd	Karen Green
Mark Schermeister	William Finegan
Eric Woodring	Randy Hickman
Jay Lefkowitz	Dean Schahrer
Nelsido Hernandez	Scott Lewis
Wendell Willis	Robin Willis
Warren A. Hragyil	Arthur Joseph
Greg Woods	Walter Deteresa
Mike Licht	Terry Culbertson
Geroge Waggoner	Leander Baltimore
Richard Kupfer	Jeffrey Warters
Pat Fier	Jerry Swords
Walter Schade	Joe Camaraza
Joseph Spencer	Scheri Tamyn
Alfonso Benitez	Sandra Hall
Bill Knost	Tom Arden
Michael Miller	Matthew Deak
Jeffery F. Schneider	Daniel Kassem
Frank Halasz	Terri Erekson
Dave Rivero	Dean Logvin
Russell Hart	Bill McDaniel
Debra Cannon	

Special Factors

- Trainers were available for 6 blocks of coaching each day—8 and 10 a.m., and 1, 3, 5, and 8 p.m.
- One PC was available for training in the simulator room. On several occasions, simulator training was being conducted at the same time as ESIS training. This was distracting, as was pointed out by several trainees.
- An 800 x 600 monitor was in place during the first session, resulting in some dialog boxes being truncated. The monitor was replaced prior to the second session.

Continued on next page

April 23, 2001

TSD Familiarization Evaluation Report - ZMA, Continued

Evaluation Feedback

The following table lists the students' evaluation of the class, based on the responses of the 43 students.

Question	Average Response
1. My ability to use TSD has increased	4.65
2. Instructors clear and understandable	4.88
3. Instructors knowledgeable	4.93
4. Training will help in my job	4.60

Scores are on a scale of 1-5:   where
1 = Strongly Disagree   2 = Disagree   3 = Neutral   4 = Agree   5 = Strongly Agree

Most valuable part of course

The students responded:
- Hands-on
- Hands On
- Hands on and explained very well
- Small group – questions answered.
- Hands on training
- How to use the TSD
- Hands-on practice of features
- Small group size made for more access to instructor, more questions + answers.
- Hands on aspect
- Hands on experience
- The weather display will be very useful – especially the lightning data.
- Hands on application
- Hands-on training (being able to ask questions)
- One on One Instruction
- Hands-on application with instructor to answer questions
- Hands on
- Hands-on experience with equipment
- Hands on
- Learning the functions
- It helped me understand TSD better
- WX info
- Hands on!
- Hands on review
- Hands-on.

Continued on next page

TSD Familiarization Evaluation Report - ZMA, Continued

Most valuable part of course (continued)

The students responded:
- Hands-on.
- Useability
- Hands on.
- The actual hands on approach.
- Face to face and hands on.
- Being able to pull up A/C
- Hands on with equipment
- Will help in CIC duties
- Hands on/interactive
- Learning to use the software for the TSD. Being able to exam flights, weather and routes together.
- Hands on training.
- Actual hands on time
- The hands-on
- Hands on training
- Info

What to include in future training?

The students responded:
- All of ESIS, not just TSD
- Detailed info on flights (lists)
- Got plenty
- WX Alerts
- Use of overhead display
- None
- Based on knowledge of system, just more time
- Just like to see some refresher and maybe some "TSD For Dummies" reference books.
- None at this time
- None
- Everything was covered.
- Any major changes
- Do not know

Continued on Next Page

TSD Familiarization Evaluation Report - ZMA, Continued

How could training be improved?

The students responded:
- Don't know
- None
- Don't Know
- …only if it was longer
- More time allotted – many questions because of system capabilities
- More time for review
- Dispense Quik Ref Cards
- Good as is. Good Pace. Good content
- Don't know
- More concentration on flights
- Don't know
- Provide objectives use ISD standards
- Not so much info in such a short period. Should be at least 8 hrs of training.
- Maybe a little more time to practice.
- Doesn't need improvement.
- More time
- Conduct in a private environment – 2 eval problems were conducted in same room!
- Larger monitor and in a room where evals are not taking place.
- More on ESIS
- More time with various commands

Other comments

The students responded:
- Sheet of TSD Menu Functions should be changed to a 3 or 4 fold pocket (shirt) sized device
- Good pace, very informative
- Great
- Excellent to have one on one instruction
- Additional training as features are added.
- More detailed instruction on flight commands
- Delete the CBI, have a position for each person to do inputs or a one on one
- No suggestions
- More hands on training with instructors.
- Good Job!

Continued on Next Page

TSD Familiarization Evaluation Report - ZMA, Continued

Other comments

The students responded:
- Time was fine
- More time
- Was an enjoyable learning exp.
- Good course that should be helpful
- Training pace good – no more than 3 students per class because number of questions grows
- I like the small class atmosphere and personal attention given by the instructor.
- I thought the class was good. It was explained in a very easy manner

Appendix F: TSD Familiarization Evaluation Reports

TSD Familiarization Evaluation Report - ZME

Dates	March 27-29, 2001
Location	ZME - Memphis ARTCC
Population	44 students
Trainers	Christine Risko and Gail Griffin

Students

David Blackwell (Supervisor)	Larry Otts (OS)
Jeff Johnston (OS)	C. M. Miller (OS)
Jim Kyle (OS)	William E. Mims
Robert Clyburn	Jason Arnold (ATCS - CPC)
John McDonald	James P. Sullins (CPC)
Allen Henninger	Mitchell Bird (CPC)
Leland Gupton (OS)	William R. Miller (CPC)
Rodney McNeill (Supervisor)	Mack Archer (SATCS)
David Dawson (OM)	George G. Simmons
Steven Kaeser (OM)	Jeff Call
Jerry P. Rippee (OS)	Charles Booker
Ed Clore (Operations Manager)	Philip Griswold
Scott Hansen	Martin W. Adams
James Studstill (ATCS/CPC)	P. Knapp (OS)
Wallace Lilly (ATCS - CIC)	Nancy Gill (OS)
Richard Anderson	Roland Hendren (OS)
Tom North (SATCS)	Ray Rowland (OS)
David Christmas	Cheryl Wright-Watkins (OS)
Chris Prince (OS)	James C. Watkins Jr. (OS)
Bucky Parsons	Robert White (SATCS)
Clint Nichols	J. C. Coughlin
Peter Ritchhart (QA)	James Score (CIC)

Special Factors

TSD Training was conducted on one workstation, which was located in the Traffic Management Office. The instructors provided training during two shifts: the first from 7:30-9:30 a.m., 9:45-11:45 a.m., and 12:30-2:30 p.m., and the second from 3:00-5:00 p.m., 5:30-7:30 p.m., and 7:45-9:45 p.m.

Clarence Goodwin, the ZME point of contact, successfully coordinated the scheduling of trainees, who were excited about learning the TSD functions. Most of the trainees were not familiar with the TSD, and a few had never seen it before.

Continued on next page

April 23, 2001

TSD Familiarization Evaluation Report - ZME, Continued

Evaluation Feedback

The following table lists the students' evaluation of the class, based on the responses of the 44 students.

Question	Average Response
1. My ability to use TSD has increased	4.60
2. Instructors clear and understandable	4.91
3. Instructors knowledgeable	4.98
4. Training will help in my job	4.49

Scores are on a scale of 1-5: where
1 = Strongly Disagree 2 = Disagree 3 = Neutral 4 = Agree 5 = Strongly Agree

Most valuable part of course

The students responded:
- The hands-on training - not just instructor demos
- TSD training
- Having the ability to predetermine traffic flow into sectors
- Hands on use
- Understanding the purpose of TSD
- Everything
- For me - a refresher of past TMU experience
- The course was good overall.
- One on one instruction
- Hands on
- Getting it
- Very thorough, several questions were asked, all completely covered.
- Increase my capacity to understand the total operation. Thanks.
- Understanding of TSD basic functions
- All of it.
- The instructor was very informative. Thanks.
- All
- All
- Ability to anticipate needs, combining, de-combining sectors, spacing of traffic
- The entire presentation was valuable.
- Hands on
- The weather information will greatly help me.
- Instruction was very helpful in demonstrating the relevance of TSD.

Continued on next page

TSD Familiarization Evaluation Report - ZME, Continued

Most valuable part of course (continued)
- Hands on aspect
- Small number of "students" allowing for more hands-on technique.
- Alerted sectors
- Using TSD in real world situations
- All
- All
- Small group setting with personal instruction
- All
- Actual on the keyboard
- Hands on presentation
- The overall training
- The wide range of info available
- Hands on use
- Good explanation of the software and how it can be used on a daily basis.
- Actually working with the different windows and menus.

What to include in future training?

The students responded:
- How to incorporate the ESIS to the overhead projectors
- Complete - no new info needed
- Follow up through CBI
- None = None needed to perform supervisor function.
- As far as I know, everything was covered that would be beneficial to me.
- Possibly a refresher class on items not frequently used.
- Shortcut features
- Don't change.
- Current training is satisfactory.
- Unknown
- None
- None
- Not sure
- None
- None that I am currently aware of.
- None
- All essential info was covered.
- Perhaps good setups for floor use.
- Sector alert background

Continued on next page

Appendix F: TSD Familiarization Evaluation Reports

TSD Familiarization Evaluation Report - ZME, Continued

What to include in future training?
(continued)

- More hands on
- More practice exercises
- Can't think of anything.

How could training be improved?

The students responded:
- More time to do hands on practice scenarios
- No improvement needed
- No improvement needed
- Allow more time for training and incorporate a break, this covers a lot of material.
- Leave as is - small hands-on groups
- Unsure - small class and hands on worked very well.
- More hands-on.
- Include more than supervisors in training.
- Individual PC nodes for students
- More training terminals or longer hands on sessions
- Reviews
- Current training is satisfactory.
- Spend more hands on time.
- No.
- More "on-hands" time
- Review in approximately two weeks
- Not sure.
- The training was adequate.
- Class time may be extended to allow for more examples.
- More time.
- Not sure it could be.
- Cannot be improved.
- Looked good to me.
- More time - Make CBI available for supervisors.
- Good instruction - would have been nice to have seen CBI first.
- Previous CBI review could have been helpful.
- Closer to obtaining equipment - Perhaps a little longer
- More hands on in depth training
- Refresher training

Continued on next page

TSD Familiarization Evaluation Report - ZME, Continued

How could training be improved? (continued)

- The only way I could see a way to improve it is to have a follow-on class about 6 months after the facility has received the TSD.

Other comments

The students responded:
- None
- No comment
- No comment
- The class could probably be a little longer, maybe a full day of training.
- Refresher and/or training on updates would be helpful when they come up.
- Great job.
- Good job, Gail.
- Chris did a very good job.
- Good job.
- More time / hands on.
- Thank you.
- Ms. Griffin did a very fine job.
- Everything was done right on par. Training was thorough, complete and to the point.
- Add "Select All" feature for selecting long lists of items.
- Nice job!
- Well paced
- Great job!
- Gail was great - very knowledgeable, enthusiastic and encouraging.
- Ms. Griffin made an excellent presentation of the material.

TSD Familiarization Training Issues - ZDC

Dates   September 13-14, 2000
Location   DC ARTCC
Population   26 students
Trainers   Tomba Kambui, Justyne Johnson, Christine Risko, Joe Jankowski

Training Issues   The following issues were encountered during the training.
- Every time the display was initialized or recalled via the Display menu, the Precipitation Intensity legend was displayed, even though it was not a default nor saved (Trainees – All)
- Tried to remove labels from 3 sectors, but one label could not be removed (Trainee – Gary Bukovskey, 9/13, 6-8 pm)
- When requesting reports, received message that not all data was available (Trainees – Steve Privott and L'Tanya Talley, 9/14, 4-6 pm)

TSD Familiarization Training Issues - ATCSCC

Dates	November 1, 2000
Location	Command Center
Population	9 students
Trainers	Tomba Kambui and Justyne Johnson

Training Issues

The following issues were encountered during the training.
- Every time the display was initialized or recalled via the Display menu, the Precipitation Intensity legend was displayed.
- We were unable to save a colors and fonts file under Display:Adapt:Colors and Fonts.
- We could save a map items file but not recall it.
- Would not allow us to create reroute: MRIS MOL J22 VXV J46 J6 LIT.

TSD Familiarization Training Issues - ZBW

Dates	November 14-16, 2000
Location	Boston ARTCC
Population	24 students
Trainers	Christine Risko and Joe Jankowski

Training Issues

The following issues were encountered during the training.
- Although the Dot icon was defined, Jets icons were displayed. (Trainees – John Fitzgerald, Dave Emmes, Keith McKinley, 11/16, 8:30-10:30 am; Dan Ireland, Jon Schippani, 11/14, 8-10 am)
- When using Show Map Items to display sector ZBW36, both ZBW 36 and ZBW 39 were displayed. (Trainees – Dan Ireland, Jon Schippani, 11/14, 8-10 am)
- Could not resize the Select Reroutes window. Because the monitor was restricted to 800x600, only one reroute was accessible in the window. (Trainees – all)
- The system would not display Examined Flights. The Alerts Report request timed out 3 times and in each instance we were auto logged out of the ETMS system. (Trainees – J. Heinz, Ron McEwen, Ron Smith, 11/14, 1-3 pm)

TSD Familiarization Training Issues - ATCSCC

Dates	November 27, 2000
Location	ATCSCC
Population	5 students
Trainers	Tomba Kambui

Training Issues

The following issues were encountered during the training.
- One student had a programming background and wanted to know what language the software was written in.
- We were unable to save a colors and fonts file under Display: Adapt: Change Colors and Fonts.
- We could save a Map Items file but not recall it.

TSD Familiarization Training Issues - ZLC

Dates	January 10-12, 2001
Location	ZLC- Salt Lake ARTCC
Population	49 students
Trainers	Joe Jankowski, Tomba Kambui, Dag Egede-Nissen

Hardware or Software difficulties experienced	The following hardware or software difficulties were experienced during training:
	
	• On the first day of training, all computer terminals had to be rebooted at 11:50 local time due to difficulties at the Command Center. Up to that time, trainers had encountered difficulties with extracting reports, such as LIST, COUNT, and ARRD. In the afternoon, reports were available, but were very slow in appearing. The following day, reports arrived quickly and appropriately.
	
	• Throughout the training period, the following discrepancies occurred:
	
	    ▪ Mnemonics (Alt key with letter) did not work.
	
	    ▪ In Alerts, the right mouse click would not display the flight icon pop-up menu with Examined Flights.
	
	    ▪ Previously saved Map Items files could not be recalled.
	
	    ▪ In the Select Flights pop up box, when a flight set was dragged to another location, a black smudge appeared in the number column. (The pixels in that area were not displaying the color of the box.) This smudge remained until the box was closed and reopened.

TSD Familiarization Training Issues ZNY

Dates January 23-25, 2001
Location ZNY- NY ARTCC
Population 30 students
Trainers Gail Griffin, Dag Egede-Nissen

Hardware or Software difficulties experienced

The following hardware or software difficulties were experienced:
- Map items cannot be saved/recalled
- Mnemonics do not work
- PC used for training crashed on three occasions. The difficulty seems to involve the specifics of hook-up to the networked node on the floor, and is probably not a system wide concern.
- Several trainees mentioned that they would like to use the function SHOW MAP ITEM without the labels displayed. The labels can be toggled off after the item is displayed, but those labels re-appear when new SHOW MAP ITEMs are entered and under a variety of update conditions.
- Two difficulties with the ALERTS display were reported.
  - Bob Giacomazzo (631 468 1403) said that when two or more Alerts Bar Charts were displayed, after a period of time one or more of the charts stopped updating. Occasionally, one bar chart stayed active, but other charts froze and had to be closed and re-opened to re-activate.
  - George Leonard (631 468 1402) reported that the TSD display occasionally froze when several Alerts Bar Charts were displayed.

Appendix G: TSD Familiarization Training Issues

TSD Familiarization Training Issues - ZID

Dates	February 6-8, 2001
Location	ZID - Indianapolis ARTCC
Population	52 students
Trainers	Joe Jankowski, Gail Griffin, Dag Egede-Nissen

Hardware or Software difficulties experienced

The following hardware or software difficulties were experienced:

- Under the Tools menu, when a user elects to view a previously saved Snapshot, a malfunction of TSD may occur. When the box containing previously saved Snapshot files appears, the user should select (highlight) the desired file by clicking on the entry, then clicking on the **OK** button to retrieve the file. However, on occasion a user may try to shortcut the **OK** button by double clicking on the file entry instead. When that entry is double clicked, the entire TSD display disappears from the screen, as if the user had chosen to quit that TSD display.

- On Thursday, Feb 8, during the 4-6 PM training session, a student was unable to access Count, List, and Area reports after pressing Enter. The dialog box that contains the Display button did not appear. The student closed TSD, re-launched, and found that that reports would now display. A message appeared stating that not all FTM data was displayed.

- The following anomaly occurred on training PC's associated with only one TSD workstation (Workstation 4). During all sessions on Feb 8, starting at 4 AM, when the user attempted to display various Pacing Airports using the **Show Map Item** feature, those pacing airports appeared on the screen. When the cursor was moved to the location of each designated airport, that feature became highlighted and could be selected. That is, each airport could be highlighted except for Cincinnati (CVG), which acted as if inert to the movement of the cursor. This condition continued even after the workstation computer and the PC were rebooted. The user changed colors and font size for pacing airports, but that had no effect on the Show Map Item CVG entry, which continued to be unresponsive to the presence of the cursor. The TSD Hotline (Chris) at Herndon attempted to replicate the anomaly, but was unable to do so. When the Cincinnati airport was displayed through the generic Map Overlays Pacing Airports entry, the airport could be highlighted correctly. The ZID System Administrator, Barbara Roddy, was made aware of this situation and will contact the Hotline to see if it can be resolved.

TSD Familiarization Training Issues - ZLA

Dates	January 10-12, 2001
Location	ZLA - Los Angeles ARTCC
Population	38 students
Trainers	Chris Risko, Charles Mohr

Hardware or Software difficulties experienced

The following hardware or software difficulties were experienced during training:

- Throughout the 3-day training period, the following discrepancies occurred:

    - Both machines experienced technical problems each session, usually in alerts or weather portion of the training, the dialog boxes would multiply on the screen filling the screen with unreadable dialog boxes. The only way to clear up the problem was for the trainers to shut down and restart TSD.

    - One of the machines was not able to run and view reports sometimes receiving a "reports not responding" error message.

- On day 3 of the training, both machines received the following error message frequently throughout each training session. "Reroute Server Not Available". When we did get to the "Reroute" portion of the training, we were able to create a reroute, but were not able to select or view previously created reroutes.

Appendix G: TSD Familiarization Training Issues

TSD Familiarization Training Issues - ZTL

Dates	February 20-22, 2001
Location	ZTL - Atlanta ARTCC
Population	45 students
Trainers	Justyne Johnson, Charles Mohr, Dag Egede-Nissen

Hardware or Software difficulties experienced

The following hardware or software difficulty was experienced:
- The user should be able to highlight map items displayed on the screen by moving the cursor to the desired map item. This technique does not always work with Jet and Victor airways. Often, those airways will have sections that seem inert to the presence of the cursor.

For example: Center the TSD display on Atlanta with a scale of 1000. Using **Show Map Item**, display Atlanta (ATL) and Nashville (BNA). Using **Map Overlays**, display all Jet Airways. With the cursor, trace the airway J45 from the top left of the screen to the bottom right, as it passes through BNA and ATL. The airway reacts to the presence of the cursor and is highlighted until immediately northwest of BNA and again from approximately two inches southeast of ATL. Between BNA and ATL, however, the airway does not react to the movement of the cursor. The airway also cannot be highlighted near the bottom of the screen.
This example can be replicated with numerous other airways.

TSD Familiarization Training Issues - ZFW

Dates	March 20-22, 2001
Location	ZFW Dallas/Ft Worth ARTCC
Population	67 students
Trainers	Gail Griffin, Charles Mohr, Dag Egede-Nissen

Hardware or Software difficulties experienced

The following hardware or software difficulties were experienced:

- The Exceed program behaved erratically throughout the week. The ETMS log-on screen did not open consistently on the PC monitor; on several occasions it opened on the secondary monitor. Furthermore, the display could not be dragged from one monitor to the other. This situation affected training, since the PCs used for training were not connected to a second monitor. When ETMS opened in the "missing" monitor, training could not proceed until the LAN Administrator attached a second monitor and somehow manipulated the ETMS window back to the PC monitor. TSD trainers adapted by not exiting Exceed; they covered opening and closing Exceed verbally. The LAN administrator is aware of this issue and is working to correct it prior to installing PCs in work areas.

- During the very last training session, at approximately 9:00 PM on Jan 22, the Reroute function failed to work properly. The Create Reroute function worked well and displayed the reroute as desired; however, the Edit Reroute dialog box failed to open the first time it was selected from the Edit menu. Upon clicking Edit Reroute again, the Edit Reroute dialog box opened, but the reroute line itself had disappeared from the Select Reroute dialog box. (The reroute was still displayed on screen, but the Select Reroute dialog box did not list the reroute.) The trainee resolved the situation by closing the TSD display and reopening it.

TSD Familiarization Training Issues - ZMA

Dates	March 20-22, 2001
Location	Miami ARTCC
Population	43 students
Trainers	Justyne Johnson and Joe Jankowski

Hardware or Software difficulties experienced

The instructors encountered the following software issues:

- The mnemonic method of opening a menu works only if the Alt key to the right of the spacebar is used; the left Alt key does not work.

- One student claims the boundaries of ZMA40 are incorrect.

TSD Familiarization Training Issues - ZME

Dates	March 27-29, 2001
Location	ZME - Memphis ARTCC
Population	44 students
Trainers	Christine Risko and Gail Griffin

Hardware or Software difficulties experienced

The instructors encountered the following software issues:

- After every user launched the TSD, the message *Reroute Server is not available* appeared. The instructor contacted the TSD Hotline in Herndon, VA, who tried to resolve the problem. The message continued to display throughout the three days of training.

- One of the students requested a way to select print in black and white without having to change colors and fonts.

TSD 2000 Trainer Certification Checklist

NAME: _____   **FINISH DATE:** _____

Background Tasks

Task	Scheduled for	Completed
Read the Introduction, Chapters 1 – 3 and 8 of the ETMS Reference Manual.		
Read the following in Chapter 4 of the ETMS Reference Manual: Display Menu; Maps Menu; Flights Menu; Alerts Menu; Weather Menu; Hide/Show Reroutes; Snapshot Commands and Command Line Command (Tools Menu).		
Read the following in Chapter 5 of the ETMS Reference Manual: Request Area; Request ARRD, Request Count, Request List.		
Read required sections of *Fundamentals of Air Traffic Control* by Michael S. Nolan.		
Read the primer, *Airspace for Everyone*.		
Complete Modules 1-10 in the TSD Tutorial.		
Complete all activities on Certification Checklist.		

Software Tasks

Task	Self Check	Certification Check
Access TSD		
Launch Exceed.		
Launch ETMS.		
Launch TSD.		

Task	Self Check	Certification Check
Window Basics		
Active windows.		
Moving windows.		
Resizing windows.		
Maximizing and minimizing windows.		
Closing windows.		
Single and double clicking.		
Menu bar and using mnemonics.		
Tear off menus.		
OK, Apply, and Cancel on dialog boxes.		
Trackball and mouse buttons.		
Display and Maps		
Show the Times window.		
Display the following overlays: ARTCCs, pacing airports with labels, boundaries, and arrival fixes with labels.		
Center the display on a BOS with a scale of 2000.		
Change the color of the arrival fixes and the font size of the pacing airport labels.		
Save this TSD display. Record the name here:		
Create a legend.		
Center the display on ORD with a scale of 1000.		
Turn off pacing airports.		
Save this TSD display. Record the name here:		
Recall the BOS 2000 display.		
Delete the ORD display.		

Appendix H: TSD 2000 Trainer Certification Checklist

Task	Self Check	Certification Check
Display jetway J42 and save.		
Display five range rings around BOS at 50-mile increments and save.		
Find the distance from BOS to ORD to ATL.		
Explain Redraw and Initialize on the Display menu.		
Flights		
Display the Flight Count window.		
Show all flights arriving in BOS as yellow planes.		
Show all flights departing BOS as magenta A (automatic) icons.		
Add a new flight set; show all flights arriving in BOS that departed JFK in cyan and display their data blocks.		
Prioritize flights so the BOS arrivals take precedence over the BOS/JFK flights.		
Reverse the action just performed.		
Copy the flight set of all flights departing BOS.		
Change the newly copied flight set to display all USAir (USA) heavies departing BOS in red.		
Save the flight set.		
Use Find Flight to display all USA flights in yellow.		
Turn off the find flight highlighting (remove flight).		
Use the Browse Flights feature to display temporarily the data block for a flight.		
Use the Flight Icon Pop-up Menu		
Display a data block for a flight.		
Toggle between Org/Dest and Route in the data block.		
Draw the flight plan route.		
Display the history for a flight.		

Task	Self Check	Certification Check
Display the Last TZ for a different flight.		
Delete an aircraft icon.		
Change the color of a flight with a displayed data block.		
Reposition the data block.		
Alerts		
Display the alerts for high sectors.		
Examine an alerted element.		
Display the alerted element's bar chart.		
Display the alerted element's alerts report.		
Weather		
Display Canadian weather.		
Display the radar tops in blue.		
Display only precipitation level 3 and above.		
Display lightning. Show the latest strikes in pink and the strikes from 45 minutes ago in green.		
Display all jet stream options.		
Reroute		
Show reroutes.		
Hide reroutes.		
Tools		
Print the display.		
Request a List.		
Request a Count.		
Request an Area.		
Request an ARRD.		
Email		

Appendix H: TSD 2000 Trainer Certification Checklist

Task	Self Check	Certification Check
Click on the ETMS icon on the Common Desktop Environment (CDE) Toolbar.		
Click on the Email icon.		
Send a message.		

Coaching Tasks

Task	Self Check	Certification Check
"Train" instructor using the Coaching Exercises.		

Appendix I: Job Aid

TSD MENU FUNCTIONS

Use the Display Menu

If you want to...	Select this menu item	Then
Show or hide the Times window	Show/Hide Times	N/A
Initialize *(return to default display)*	Adapt	Select Initialize
Add a legend to the screen	Legend	Input the desired legend
Remove spurious graphics	Redraw	N/A
Change color or fonts	Adapt	Select Change Colors & Fonts
Save your display	Adapt	Select Save
Recall your display	Adapt	Select Recall

Use the Maps Menu

If you want to...	Select this menu item	Then
Center the display	Move/Zoom*	Input a lat/lon or point
Zoom	Move/Zoom*	Input the desired width
	* or place cursor at desired center point and type M to center, Z to center & zoom, U to center & unzoom, or X to undo last move/zoom command	
Display an individual map item	Show Map Item	Input the designator
Display range rings	Range Rings	Input the information
Find distance between points	DME	Click on locations
Display overlays *(sectors, fixes, etc.)* and their labels	Overlays	Select appropriate checkboxes

Use the Flights Menu

If you want to...	Select this menu item	Then
Show or hide selected flights	Show/Hide Flights	None
Show or hide flight counts	Show/Hide Flight Counts	None
Change the flight icons	Select Flights	Select a different icon
Add a new flight set	Select Flights	Select Flight Set menu, New
Copy a flight set	Select Flights	Select Flight Set menu, Copy
Delete a flight set	Select Flights	Select Flight Set menu, Delete
Move a flight set to the top of the list	Select Flights	Highlight the row, then select Flight Set menu, Top
Save your flight sets	Select Flights	Select File menu, Save
Find a flight	Find Flight	Input the ACID

Note: To filter by aircraft type codes, use **H** *(heavy)*, **J** *(jet)*, **P** *(prop)*, **T** *(turbo-prop)*

Use the Alerts Menu

If you want to...	Select this menu item	Then
Show or hide selected alerts	Show/Hide Alerts	N/A
Select an alert	Select Alerts	Select the type
Display an alert timeline	Examine Alerts	Input the type and name
Display a bar chart	Examine Alerts	Select Alert menu, Show Bar Chart
Display an alert report	Examine Alerts	Select Alert menu, Alerts Report

Use the Weather Menu

If you want to...	Select this menu item	Then
Show or hide weather	Show/Hide Weather	N/A
Select the weather to display	Select Weather	Choose the desired weather
Select the precipitation levels to display	Select Weather	Select NOWRAD Legend, then click desired level
Display Lightning, Tops, and/or the Jet Stream	Select Weather	Choose the desired option(s)

Appendix I: Job Aid   Page I-2

TSD MENU FUNCTIONS, *continued*

Use the Reroute Menu

If you want to...	Select this menu item	Then
Show or hide all reroutes	Show/Hide Reroutes	N/A
Show or hide individual reroutes	Select Reroutes	Click the appropriate Show box
Create a reroute	Create Reroute	Choose the desired options
Edit or delete a reroute	Select Reroutes	Choose the Edit menu command in the Select Reroutes dialog box

Use the Tools Menu

If you want to...	Select this menu item	Then
Print a display	Snapshot	Select Snapshot, Capture
Request a report	Command Line	Input the request

Other Functions

Accessing TSD
Double-click the Exceed icon.
Log into ETMS.
On the ETMS toolbar, click the up arrow above the ETMS icon.
Click on TSD.

Mouse Buttons
Left Button – to click and double-click buttons, fields, commands, icons, check boxes, etc.
Right Button – to display pop-up menus.
Middle Button – to drag data blocks, reroute names and flight sets, etc. Use the right button on a two button mouse.

Bar Chart
Bar Chart Red – the number of airborne flights.
Bar Chart Yellow – the number of flights still on the ground.
Note: The capacity level is set by the controlling agency and input by the TMU.

Exit Steps
Close TSD – click on Display, Quit.
Close ETMS – click on Exit icon.
Close Exceed – press Logout: Save Current button.

Documentation
The ETMS Reference Manual and Tutorial are available in the TMU and on the Command Center's website.

OK, Apply, Cancel, and Help in Dialog Boxes
OK – changes take effect and the dialog box closes.
Apply – changes take effect and the dialog box remains open.
Cancel – changes do not take effect and the dialog box closes.
Help – help opens

Sector Gridlines and Colors
\\\\\\   low
//////   high
|||||| superhigh
Red – The flight that caused the alert is airborne.
Yellow – The flight that caused the alert is not yet airborne.
Green – The controllers for that sector, fix, or airport told the TMU that they could handle the flights.
Purple – The sector, fix, or airport is not alerted but is being examined.

Herndon, VA
7 a.m. – 7 p.m. ET
703-904-4434

TSD Hotline Numbers

Volpe
7 p.m. – 7 a.m. ET
617-494-2556 or 617-494-2557